GHOSTS BEHIND GLASS

DOLLY JØRGENSEN

Ghosts Behind Glass

ENCOUNTERING
EXTINCTION
in MUSEUMS

The University of Chicago Press   CHICAGO AND LONDON

The University of Chicago Press, Chicago 60637
The University of Chicago Press, Ltd., London
© 2025 by The University of Chicago
Published 2025
Printed in Türkiye

34 33 32 31 30 29 28 27 26 25   1 2 3 4 5

ISBN-13: 978-0-226-84264-6 (cloth)
ISBN-13: 978-0-226-84230-1 (paper)
ISBN-13: 978-0-226-84231-8 (ebook)
DOI: https://doi.org/10.7208/chicago/9780226842318.001.0001

Library of Congress Cataloging-in-Publication Data

Names: Jørgensen, Dolly, 1972– author.
Title: Ghosts behind glass : encountering extinction in museums / Dolly
    Jørgensen.
Description: Chicago : The University of Chicago Press, 2025. | Includes
    bibliographical references and index.
Identifiers: LCCN 2024058193 (print) | LCCN 2024058194 (ebook) | ISBN
    9780226842646 (cloth) | ISBN 9780226842301 (paperback) | ISBN
    9780226842318 (ebook)
Subjects: LCSH: Extinct animals—Museums. | Natural history museums. |
    Extinction (Biology)
Classification: LCC QH88 .J29 2025 (print) | LCC QH88 (ebook) | DDC
    591.68—dc23/eng/20250123
LC record available at https://lccn.loc.gov/2024058193
LC ebook record available at https://lccn.loc.gov/2024058194

♾ This paper meets the requirements of ANSI/NISO Z39.48-1992
(Permanence of Paper).

for MY FAMILY,
who visited so many of
these museums with me

Contents

Illustrations

All photographs are by the author unless otherwise indicated.

Ghost Stories

This book is about ghosts. In the halls of natural history museums all over the world, you peer through the glass and come face-to-face with ghosts. I am not being metaphorical, and I don't just mean something dead. I mean real ghosts—presences that haunt us. Being dead is not the end of existence if we let the dead still live and act in the world. Behind museum glass, the extinct dead are not just remembered; they live on as other presences.

They may even proliferate after extinction as they are *re-presented* through the stories told about them and their remains. This re-presenting, which means helping something to be present again through their life history and body, allows the extinct in this book, these ghosts, to act. They speak their own stories underneath the stories spoken for them. They ask for emotional and intellectual engagement. Their stories can resonate even today.

To respect the liveliness of the extinct dead is to move their stories forward. Lively relations are shared connections that may exist even after biological death. Just as we still have memories and linkages with departed loved ones, the relations with extinct beings live on in other forms. Standing in front of the glass seeing an extinct animal is standing in the presence of something that refuses to disappear. Come along with me into the museum, peer through the glass, read their ghostly stories, and allow the extinct to be present again.

Canard colvert

Coq bankiva

Great auk inside a
glass case, Muséum
d'Histoire naturelle,
Nantes, France, 2024.

Grand pingouin

GHOSTS BEHIND GLASS

1

Spectral Encounters

In the Naturalis museum in Leiden, the Netherlands, a whole gallery is dedicated to "Death." The visitor journeys through accidental animal death, decomposition, and the practice of taxidermy in the eerie, dramatically lit space. Death feels all around. Toward the end, there is a small display titled "Uitcestorven / Extinction." After all, what could be more dead than something that is extinct?

In the Extinction display, on the left, there is a small dinosaur toy. Dinosaurs are the ultimate symbol of extinction. While you can see dinosaurs in films and books, the natural history museum is the place where people go to see dinosaurs. Dinosaurs, as "gigantic terrible lizards" (this is what the Greek roots of the word *dinosaur* mean), captured the popular imagination in the Victorian age through displays like the Crystal Palace Park dinosaurs in 1854. Dinosaurs became spectacles for the public as their remains were put on display in museums.

FIG. 1 (*facing*)
The dino and the dodo, Naturalis museum, Leiden, the Netherlands, 2022.

And visitors have flocked to see them ever since. Dinosaurs are fascinating because while these giants once ruled the planet, they are all gone. When you ask someone to name something extinct, they are probably going to name dinosaurs first.

On the right is a dodo model. This dodo stands in for the modern extinct animals—passenger pigeons, thylacines, great auks, Steller's sea cows, Japanese river otters, and hundreds more—that also inhabit the halls of natural history museums. Unlike the dinosaurs, which died out millions of years ago, these have all died out in the last 1,000 years. And, unlike the dinosaurs and all the species on Earth that died out in the previous five mass extinction events, modern extinctions have happened because of humans. In the extinction display in Leiden, the names of these modern extinct animals fill the wall behind the little toy dinosaur.

Humans have a penchant for modifying their surroundings to make the land more habitable for them. They chop down forests, drain wetlands, dam rivers, dig minerals, and raise cities. All these activities may be great for humans, but they aren't so great for the animals that used to live in those places. Human activity on Earth is radically changing the planet, particularly the use of fossil fuels, which is altering the climate. On top of the large-scale consequences of human activity, humans hunt specific animals—for food, for goods, for fashion, for medicine, for pest control, and even for collections. It is no wonder that species are going extinct at a rate that far exceeds the expected number of extinctions. We know that most species naturally go extinct over time—they either fail to thrive or evolve into something else—and this is a perfectly normal part of life on Earth. But every once in a while, a mass extinction event happens. The one that killed off the dinosaurs (or at least caused the remaining dinosaurs to evolve into birds) at the end of the Cretaceous period 65 million years ago is probably the most widely known, but there were four others before that in which the vast majority of species on the planet were wiped out. Now scientists think we are entering a new sixth mass extinction. Modern extinction, called *Anthropocene defaunation* by Rodolpho Dirzo and colleagues, is pervasive: over 320 known vertebrate species have become extinct since 1500, and the remaining species have an average 25 percent decline in numbers.

Extinction histories are traumatic histories. Extinction is an act of violence to reciprocal relations between humans and nonhumans. This violence is not evenly distributed—species in isolated ecosystems like islands are easily eradicated, as are species in areas of settler colonial development—and it is not inevitable. It is not a new revelation that humans are causing rapid extinction. Writing in 1891, Fredrick Lucas, a scientist with the Smithsonian's natural history collection, observed,

> It is not, perhaps, generally realized how extensive and how rapid are the changes that are taking place in almost the entire fauna of the world through the agency of man. Of course changes have perpetually taken place in the past through the operation of natural causes, and race after race of animals has disappeared from the globe, but there is this wide difference between the methods of nature and man; that the extermination of species by nature is ordinarily slow, and the place of one is taken by another, while the destruction wrought by man is rapid, and the gaps he creates remain unfilled.

The timing of these modern extinctions has meant they are documented and presented to the public as a contemporary event, unlike the extinction of the dinosaurs, which died out millions of years ago. This can mean that the species on display are remembered by individuals or cultures, or even that the classification of species on display might need to shift from extant (living) to extinct. Because humans are implicated as the cause of modern extinctions, the narratives of species loss are entangled with human histories. To encounter extinction ghosts, you need to visit their modern sepulcher: the natural history museum. It is here that you encounter them through their bodies and the stories told about them.

Natural history museums trace their roots to early modern cabinets of curiosities in which the wealthy collected and displayed the wonders of nature. A famous engraving of Danish naturalist Ole Worm's cabinet of curiosities from 1655 includes a room full of seashells, cultural artifacts, gemstones, and stuffed specimens hanging from the ceiling. On one shelf, there is even a great auk, a species that is now extinct. By the nineteenth century, Europeans were enamored with natural history, believing that it was good moral education to revel in God's creation. Public museums based on personal collections, like Sir Hans Sloane's collection, which was the kernel of the Natural History Museum in London, were built because they appealed to elite sensibilities about education for the masses. Already in 1773, before the United States had even declared its independence, a natural history museum was established in Charleston, South Carolina, to present a "full and accurate Natural History" of the area. National natural history museums grew up with the nation-state. For example, the Muséum national d'Histoire naturelle in Paris, founded by decree in 1793, was a product of the French Revolution to advance the revolutionary agenda to popularize scientific thinking.

These early natural history museums collected rare specimens, even sending out missions to kill and collect the last known individuals if the species was thought doomed. For example, in 1886, the Smithsonian in the United States supported an expedition to collect American bison, which were on the brink of extinction. William Hornaday, who led the expedition, was afraid that "by the time the museum-builders of the world awake to the necessity of securing good specimens of all these [referring to a list including ones that become extinct] it may be too late to find them"; therefore, he urged immediate collecting expeditions. Historian Hanna Rose Shell concluded that Hornaday "embraced the notion that he himself, on behalf of the National Museum, should kill some of the last wild buffalo in order to save, which is to say embody, its memory in corporeal form." The museum collected animals to keep memories of them from fading away. In 1915, Oliver Farringdon, curator of geology and president of the American Association of Museums, observed:

A large share of the animals and plants inhabiting this continent [North America] at the time of its discovery by Europeans are not destined to survive long except as they are protected by man, and some will become extinct in spite of him. . . . To museums must be largely assigned the work of conserving the remains of such forms ere they are absolutely lost. Specimens which are valuable now will be priceless in the years to come.

The thought was that it was better for the museum to collect specimens while they could than to let the species disappear into oblivion. The museum was interested in keeping extinct species alive even in death.

The remains of extinct (and rare) animals (whether they are eggs, nests, bones, or skins) have generally ended up in natural history museum collections because they are valuable to science as well as curiosities and wonders of nature. But very few of these are visible to museum visitors—most of the valuable specimens that museums hold are locked away in their back rooms. Thousands upon thousands of animal bodies and parts sit in drawers and cabinets, normally accessed only by scientists. Some writers have gone to the back rooms and seen many rare and wonderful specimens, as Michael Blencowe did for his book *Gone* or as Miranda Cichy did for her inspirational article on encounters with paradise parrots. But this book is not about what museums hold behind the scenes. This book instead examines how extinction is presented to the public through exhibitions. This is an exploration of what the visitor who pays for a ticket and walks the halls sees when looking at the displays and reading the labels. (It is a myth that people don't read labels; in my visits to natural history museums, I see that people are always reading labels.) Together we will search the display cases for extinct ghosts, their remains, and their stories.

I visited over seventy museums across the globe to encounter extinction. I made trips to visit the famous natural history collections, like the Natural History Museum in London, the Smithsonian in Washington, DC, and the Naturhistorisches Museum Wien in Vienna. I specifically went to see museums that had extensive displays of a particular extinct animal, like the Mauritius Natural History Museum and the Tasmanian Museum and Art Gallery. But I also visited many small museums, often stopping in because I was in the area for something else (such combinations were critical for reducing the environmental cost of these visits). On a stopover of a few hours at the Frankfurt airport, I took a taxi over to the Naturmuseum Senckenberg for a visit before zipping back to the airport to board my plane. Nearly all the places I visited were natural history museums or museums with a significant natural history collection, although there are a few exceptions, like a special exhibit I saw at the British Library. On some occasions I went back to the same museum again, allowing me to see changes to exhibitions. The appendix lists the details of what extinct specimens were on display in which museums.

The extent of these visits means that unlike a detailed study of one extinction exhibit, this book is able to identify trends, similarities, and differences in display techniques and modes of storytelling across the world. In that, this book is a guide for future visitors (and even museum curators) when they encounter extinction. It helps identify how a given exhibition fits or doesn't fit into larger patterns. How is extinction presented through different exhibition techniques? What stories are told with the bodies on display? What happens when you are face-to-face with an extinction ghost?

FORM AND FUNCTION

In order to make sense of the exhibitions discussed in the rest of this book, I will first outline the various possibilities for putting extinction on display in a museum. The form of the remains and the grouping into which they are placed allow the extinct animal to communicate in different ways. Ghosts in the natural history museum are invoked through various forms. Taxidermy, study skin, skeleton, model, and image are the most common forms exhibited, each of them with affordances, limitations, and histories. Rather than focus on taxidermy as the only mode of physical presence, this book includes many types of being, from taxidermy to model to skeleton to egg to artwork.

Encountering a taxidermy mount, a pelt, and two skeletons of the same species side by side is a rare occurrence, but it happens in the thylacine room in the Tasmanian Museum and Art Gallery. Each of these forms gives different insight into the animal. The stuffed thylacine may look most realistic because it is a three-dimensional representation of the body, but the pelt has much-better-preserved color and striping. The skeletons look the most dead, which contrasts with the nearby film of a thylacine walking in a cage looking very much alive. This room is a good entry point for considering how those different animal remains portray the ghosts of animals past.

When most people think of a natural history museum, they probably imagine either a room full of dinosaur bones or a room crammed with stuffed animals. There's good reason for the latter, as taxidermy is the primary mode of presentation for mammals and birds. Taxidermy is the art of creating a lifelike sculpture of an animal from its skin. The word derives from the Greek words "taxis" (arrangement) and "dermis" (skin). To create a taxidermied specimen, the animal first needs to be skinned to remove all its inside material. The skin needs to be treated with chemical preservatives (taxidermists in the past used arsenic for this) in order to keep it from degrading quickly. Generally, most or all of the bones are removed and a wire or plaster frame is created to act as the skeleton of the new taxidermy animal. Taxidermists choose the position to use for the mount. Perhaps the animal should be lying down or walking or stretching or pouncing—all of these are possible by creating a frame in that position. After stretching the skin onto the frame, it is stuffed with filling material and sewn together. That's why

you can refer to a taxidermied specimen as a mount (it is a skin mounted on a frame) or as a stuffed animal (it is skin that has been stuffed). Glass eyes are placed into the eye sockets since eyeballs cannot be preserved with the skin, and other soft material like nose tips might also be replaced. A taxidermied animal is "at once lifelike yet dead, both a human-made *representation* of a species and a *presentation* of a particular animal's skin," as observed by taxidermy historian Rachel Poliquin.

Taxidermy has been practiced since the early days of the cabinets of curiosities, but practices gradually became more standardized through major taxidermy workshops and competitions for the best-prepared taxidermy specimens in the late nineteenth century. The use of taxidermy for natural history specimens spread globally at that time. Liu-Chuan Tai has traced the rise of taxidermy in the Shanghai Museum, and she quotes an anonymous report from an 1875 newspaper that captures the wonder of taxidermy for the audience:

> There are small furry critters such as marshland and lakeside boars and rodents; flying animals such as eagles, magpies, and sparrows; and countless aquatic animals. There are also praying mantises, grasshoppers, and butterflies. Each is clearly marked with its name and type, as well as a location and the name of the donor. For both Chinese and foreign visitors, admission is free. The Westerners have clever techniques to keep the feathers and fur intact without falling off of the skin. Chinese have now also learned these techniques.

Taxidermists radically shaped the natural history museum's presentation of animals. As writer Mary Anne Andrei has shown, businesses like Ward's Natural Science Establishment trained taxidermists in the art and science of preparing specimens from the 1870s and, in turn, revolutionary exhibitions that displayed animals in motion and interacting with others in scenes began to appear in natural history museums. Taxidermy turned dead animals into lively ones.

Taxidermy plays with and bends time by making the dead look alive. At the same time, while taxidermy appears to make time stop for a specimen, time actually does go on. Specimens continue to degrade over time because of their organic material nature. Museums do everything in their power to keep that from happening: they often keep specimens in climate-controlled environments with constant temperature and humidity, apply pest control, repair or refurbish specimens, and keep specimens out of harmful light rays. A museum is always fighting against the breakdown of the dead bodies.

But taxidermy is not the only way to prepare museum specimens. In fact, far more specimens in museum collections are study skins. These are the skins from animals that have been preserved and often lightly stuffed but not mounted. Sometimes there are a few bones left with the skin. Sometimes the skull is still in place (imagine a hunter's bear rug on the floor). Study skins take up much less room because they can be put into flat drawers. As the name implies, they are mostly used for scientific study, but sometimes they are displayed to visitors.

We can take a look at the extinct paradise parrot (*Psephotellus pulcherrimus*) in two UK museums to get a feel for the difference in presenting a taxidermy mount versus a study skin. The paradise parrot was native to eastern Australia and was last seen in 1927. At Wollaton Hall in Nottingham, the paradise parrot is mounted in a flying position with wings outstretched. Glass eyes have been placed into the eye sockets to give it a feeling of life. At Manchester Museum, the paradise parrot is a slightly stuffed skin shown flat, belly-up. The feet are tied together and the specimen is pinned to the board. These displays give off a very different vibe. The Wollaton parrot is quite lively looking; Manchester's is very obviously dead. A big part of that has to do with the positioning of the animals (flying versus belly-up; free versus tied), but also the eyes (glass eyes versus empty sockets).

FIGS. 3 & 4
Comparison of paradise
parrot displays. *On
left*: Wollaton Hall,
Nottingham, UK, 2022.
On right: paradise parrot
with South Island Piopio,
Manchester Museum,
Manchester, UK, 2019.

The two preparations evince a different emotional response
in the viewer. Poliquin has described the physical encounter
between viewer and taxidermy as the *experiential narrative*,
which she defines as "a recognition of the embodied 'thingness'
of displays of mounted animals: the strange aura of lively yet

dead creatures collected together for the purpose of looking."
While she was discussing specifically taxidermy, the idea of
experiential narrative would apply to other remains as well,
but the experiences are different when the animal remains
are prepared differently. The study skin has the strange aura

of having once been alive and yet now dead while the viewer looks at it.

Bones are another material used in natural history displays. These bones can be loose, for example a skull may be displayed alone, or they can be articulated into skeletons. When made into skeletons, the body position is the choice of the preparer, so just like taxidermy, skeletons can show motion if desired.

Two extinct Honshū wolves (*Canis lupus hodophilax* or *Canis hodophilax*) in the National Museum of Nature and Science, Tokyo, demonstrate the difference between presenting bones versus taxidermy. The taxidermy Honshū wolf appears in the "Animals of the Earth" gallery, which features taxidermy animals from across the globe in one room. This particular individual had been exhibited alive at the Ueno Zoological Garden in the late 1800s and then shown as a stuffed specimen in the zoo's museum after death. It was later moved to the National Museum. The skeletal wolf is in an "Endangered Organisms" display in a gallery about the Japanese archipelago environment. While both exhibitions indicate that the wolf is extinct (the little owl pointing at the wolf says, "This is the Japanese wolf that is extinct"), the skeleton gives the feeling of the animal being dead for a long time. We would expect an animal to degrade and lose its skin first, with bones surviving a long time afterward. Yet these specimens are from similar time frames close to the extinction of the wolf in 1905.

Even having one bone on display can give a signal to the viewer about the size of the animal. Take displays of the giant moa, for example. The giant moas (*Dinornis*) were a group of giant flightless birds that lived in New Zealand. There were separate species on the North Island (*D. novaezealandiae*) and South Island (*D. robustus*), but both died out by 1500 due to

FIGS. 5 & 6 (*facing*)
Two presentations
of the Honshū wolf.
National Museum of
Nature and Science,
Tokyo, Japan, 2023.

FIGS. 7 & 8 Comparison of displaying a giant moa leg at the University Museum of Zoology, Cambridge, UK, in 2022 versus a whole giant moa skeleton in the Harvard Museum of Natural History, Cambridge, Massachusetts, USA, in 2023.

human predation and land clearance by the Indigenous Māori. Moa are often shown using just the leg bones. Many of these bones are models based on bones held elsewhere. These are intended to demonstrate the size of the animal by having the visitor see the leg bone and think about their own height. Sometimes, like at the Harvard Museum of Natural History in Cambridge, Massachusetts, a full moa skeleton is on display. This specimen towers over the viewer and leaves no doubt about its gigantic size.

In addition to bones, eggs are sometimes displayed to represent extinct birds. The most common are the elephant bird and the moa because of their impressive size. Great auk eggs are in several museums because of their unique patterning. Oology, the science of egg collecting, was very popular in the nineteenth century, so many museums acquired eggs of species that would later become extinct. Eggs are often displayed as aesthetic objects because of the colors, patterns, and even shape of the eggs of different species. Eggs do not reveal the contents inside, so other visual clues are required to connect the egg to a specific bird such as drawings of the parent. Eggs, however, do have an association with life in many cultures. This means that although both bones and eggs are calcium remains, they can elicit different reactions from a viewer.

Once a species is extinct, there is no possibility of getting more specimens to create taxidermy because this requires a newly dead individual. Bones and eggs are sometimes found after extinction, but they too have a short existence unless they become fossilized. Thus, there are a limited number of extinct specimens available for museums. But even when a museum does not have organic animal remains to display, they can still display extinction through models of taxidermy, skeletons, and eggs. The use of models is extremely common with some species like the dodo, which has no preserved taxidermied specimens because the remains collected when it was alive have all degraded or been destroyed. These models allow museums to display a whole animal without having organic materials and to show visitors what the animal would have (or might have) looked like when alive. In addition, skeletons are often incomplete, so model bones may be created to fill in the gaps if the curator wants to show an articulated skeleton. There is flexibility in the materials curators choose and "imperfect" specimens can be touched up or fixed. Sometimes models or replacement parts are labeled as such by the museum; at other times, they are not.

Models, however, are not just "faithful" re-creations of specimens. Sometimes, only skeletons remain of a species (like the dodo and the Steller's sea cow), so reconstructions of the bodies require some guesswork. In addition, scientifically accepted models can shift over time. The classic case of this is the dodo model. In some museums, the dodo label will note that the specimen is out-of-date because of newer research on dodo skeletons that has reinterpreted the body configuration. Some museums have even commissioned new dodo models to replace their older ones. Interpretations are inherent in models, thus they have to be understood as art as much as science. For example, the Blaschka brothers' glass models of undersea creatures are renowned for their beautiful aesthetics as much as scientific accuracy.

In addition to physical remains and models, we can encounter images of an extinct species in museums. Sometimes these are artistic works that were created while the animal was still living, like John James Audubon's mid-nineteenth-century drawings of several birds that became extinct in the twentieth century, but many are artistic impressions created after extinction. In some cases, the animal died recently enough that there are photographs or other recordings that the artist can use to inform their drawings or paintings. Other times, artists may use the remains in museums to create their pictures. Images sometimes complement the organic or model remains in the cabinet, although they can also appear on their own. In either case, images, just like taxidermy and models, are created and do not objectively represent the world.

For example, the Academy of Natural Sciences in Philadelphia has three murals of extinct birds that line the walls near two extinct bird dioramas with taxidermied specimens. The murals, painted between 1941 and 1943, juxtapose ancient extinct birds with modern extinctions. The first mural encountered when entering the hall shows the Late Jurassic archaeopteryx, one of the earliest birds, in a dinosaur-inhabited palm tree landscape. The second is a scene on a beach with three giant moa—one towers over the viewer in the foreground—and lush forested mountains in the distance. The third mural is "Birds of the Mascarene Islands," which shows thirteen species of extinct birds (plus the non-extinct pink pigeon) from the islands of Mauritius, Reunion, Rodrigues, and smaller islands in the archipelago. The dodo is in the center of the image surrounded by the other lesser-known (and unnamed in the image or label) extinct species. The Mascarene Islands mural summarizes the great loss of birds in the archipelago in one impossible view with the birds coexisting in the same habitat. While some of the birds might have interacted in life, many never would have because their preferred habitats and island homes were spread hundreds of miles apart. Extinction can

facilitate the creation of new groupings of species that were previously separate. As a whole, the Academy's artistic large-scale visions of extinct birds invite the viewer into lost worlds that are products of imagination and artistic license.

All these animal remains—whether they are stuffed animals, bones, eggs, models, or art—offer only small glimpses into what the animal once was. None of them represents the whole animal, but they do come to stand in for them and can re-present them. Literature scholars Sarah Bezan and Robert McKay have observed that "animal remains refuse to be remaindered away" because they are both ubiquitous and diverse, from fossils to taxidermy to art. Animal remains remain. They are in a state of continued presence despite the whole living animal no longer being. All the remains, regardless of type, invoke the presences of the extinct ghosts behind the glass. They imply an animal that once lived and breathed without being that animal. They are not the full corporal being, but the shadows or remains of it.

FIG. 9
Facing the mural of the giant moa, with the archaeopteryx mural on the right and the passenger pigeon diorama on the left, Academy of Natural Sciences, Philadelphia, USA, 2023.

Extinct remains appear in various contexts within a museum. In permanent exhibitions, extinct species may be grouped together into one display about extinction or they may be distributed throughout the museum with either their respective geographies (Australia, North America, etc.) or animal type (bird, mammal, reptile, etc.). They may also appear in special exhibitions that have entirely other ways of ordering the specimens.

Extinct animal remains are placed into the museum galleries based on Western scientific ideas of the ordering of species. Attempts to systematize nature grew hand in hand with the development of the early modern natural history cabinets. In the early eighteenth century, naturalist Carl Linnaeus developed a system to name species and relate them to one another, so it is not surprising that when natural history collections opened up for the public to visit in the nineteenth century, they often showed animals in groupings of related animals. Charles Darwin's theory of evolution encouraged such groupings to show how features evolved. Because natural history museums were designed in large part for education, they have generally deployed strict systematization in their displays in order to communicate contemporary scientific theories about life on Earth. Natural history groupings reflect these historical developments.

The extinction-themed cabinet is the most common option for displaying extinct specimens in natural history museums. In this arrangement, all the extinct species are placed together in one case since they all share the "extinct" quality. Sometimes this display also includes endangered animals in order to broaden it. A good example of this is the Extinct and

Endangered Animals display at the Museum of Natural History at Wrocław University, Poland, which includes numerous taxidermied specimens, an egg, an unmounted skin, and a work of art. The animals selected include extinct ones like the Javan lapwing, Carolina parakeet, and passenger pigeon as well as endangered species like the kakapo and tuatara. The most extensive example of this approach is the Muséum national d'Histoire naturelle in Paris, which dedicates an entire gallery to endangered and extinct species.

The advantage for the museum in this kind of case is that general information about the threats to wildlife, including extinction as the ultimate threat, can be shared once rather than many times throughout the museum. Many extinction cabinets will include an information board that presents the loss of species in the modern era as a trend because of hunting, habitat loss, climate change, pollution, and other causes. A mark against the "extinct" grouping is that it divorces the species from where it lived or what kind of life it had. The passenger pigeon, great auk, and thylacine have nothing in common except that they are all extinct.

Instead of the "extinct case" organizing principle, some museums place their extinct specimens with their relative animals. For example, at the University Museum of Zoology in Cambridge, UK, a pair of huia are in the display along with other birds. In this display, each bird has its conservation status as determined by the International Union for the Conservation of Nature (IUCN) in the colored part of the label on the left. The huia status is extinct, which is why the label's circle is black. Choosing to display the birds together has the advantage of showing the visitor what type of animal the extinct one is through association. You can see all the ones that are not

ZWIERZĘTA

WY

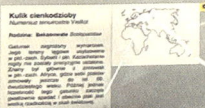

ARŁ I GINĄCE

FIG. 10
The Extinct and
Endangered Animals
display, Museum of
Natural History at
Wrocław University,
Wrocław, Poland, 2021.

extinct in addition to the ones that are. This could potentially lead to thinking about existing species as ones that could be spared extinction.

A third alternative is to place the extinct animals into their geography. Such is the case with the passenger pigeons in the American Museum of Natural History, which are placed into the gallery of New York animals. This placement puts them alongside other animals that inhabit (or inhabited) the state. Another example is the thylacine in the Bristol Museum and Art Gallery, where it is placed in a cabinet with other Australian animals. Curators making this choice are relating the species to its former geography (its home) and other species that it might have interacted with in those geographies.

An extinct species can also be placed in a thematic gallery. For example, the Finnish Museum of Natural History's specimen of the Steller's sea cow, which is the most complete skeleton in the world of this animal, is shown within the context of a comparative anatomy display. Here, skeletons of various species are put side by side to present how bones function and how scientists can use bones to identify species. Naturalis in Leiden has placed a model dodo in its thematic exhibition *Death*, which deals with various ways that animals die and how bodies decompose, as the representative, along with a plastic dinosaur toy, of extinction as the ultimate death.

Each of these groupings changes how the visitor encounters the extinct animal. When the animal is in an extinction display, its status as extinct is most clearly communicated to the visitor. This is what becomes the central characteristic of the species. When shown with either the type or geographical grouping, the animal is placed into an evolutionary context of relations. The visitor to Cambridge sees the huia primarily as

FIG. 11 (*facing*) Huia displayed with other birds, University Museum of Zoology, Cambridge, UK, 2022.

FIG. 12
The Steller's sea cow
is shown within a
comparative anatomy
display, Finnish
Museum of Natural
History, Helsinki,
Finland, 2022.

a bird, rather than an extinct thing. When it is shown in other types of groupings, like the comparative anatomy display, the extinct specimen is just one of many data points for studying the phenomenon in question. There is no right or wrong in these display choices, but they create different connections between species and environments.

GHOST STORIES

Labels and information boards matter in telling the stories of extinct specimens. Natural history museums from the start had educational aspirations, wanting to impart knowledge to visitors, with the intent of making them better-informed citizens. Historically, labels were written in authoritative tones and with specialized language; in the case of the natural history museum, this means using biological terms like species, gestation, and range. Many museums began to shift their labeling techniques in the 1980s. Labels became much shorter and are often enhanced with large, bold fonts, images, and even dynamic audiovisual content to make them pop for the viewer. These types of labels stress dialogue with the audience, often asking questions and prompting reflection. Contemporary interpretative labels often relate content based on the curatorial goal of the exhibit; they don't try to include every fact or story.

How an extinct species is labeled is key to visitor encounters because it is often not self-evident that the animal is no longer alive somewhere out there. Extinct specimen labels in some museums are quite minimal: scientific name, common name in the language of the museum, and an indication of the extinct status. But most share some kind of information about the creature's extinction history, like where the species lived, when the last one died, when the last one was seen, and the reasons it became extinct. This is succinctly written in most cases—a couple of sentences is typical. If the specimens are in an extinction-themed vitrine, there is also typically a separate longer informational board that will discuss extinction as a contemporary phenomenon.

In almost all natural history museums, extinct species are labeled as extinct, although sometimes the uncertainty of extinction affects the label's contents. The extinct indication can take different forms, commonly an IUCN label or the word *Extinct* (of course, this will be in the museum's chosen language or languages). Naturmuseum Senckenberg in Frankfurt made

FIG. 13
Carolina parakeet
displayed with a label
including an extinction
X over the map instead
of marking its range on
the map, Naturmuseum
Senckenberg, Frankfurt,
Germany, 2022.

the choice to emphasize the extinct status and placed a big red
X over a global map. While the impetus to stress the extinction
is understandable, such a choice also divorces the species from
its prior habitat, which is not marked on the map—the species'
history is erased; it becomes a ghost without an original home.

Both labels for the individual animal specimen and signage in the case or gallery help (or potentially hinder) communication of the extinction story. It can add the context of the loss of the animal, denying a complete erasure of its prior existence.

THROUGH THE GLASS

The National Museum of Ireland's natural history division has been nicknamed "the dead zoo," but the natural history museum is not really like a zoo. Large zoo animals are typically quite far away from the spectator and are presented in modern zoos without visual barriers between the visitor and inhabitant. Most natural history museum displays are exactly the opposite—the visitor is right next to the specimen, but there is a physical barrier of glass between them.

You aren't supposed to think about glass. It is one of those materials that is supposed to be invisible. The glass serves as a transparent medium for letting both light and the human gaze onto the object while protecting it from the atmosphere, potential pests, and pestering hands. Glass cases have a specific history in the natural history museum. Whereas visitors to early modern cabinets of curiosity were invited to touch and smell objects, in the nineteenth century there was a shift to sight as the primary sense and more control over the objects, especially as the audiences broadened from the elite to all members of the public. As Brita Brenna found in her study of the glass cases in the Bergen Museum, glass facilitated a division of the collections—one for public exhibition with glass and the other for scientists that allowed handling of the objects. For the public, the glass case lends an air of credibility and importance to things behind it, conferring an aura of preciousness.

FIG. 14
A great auk specimen
behind two glass
layers—the glass of
the main case and a
separate glass box,
Museo Civico di Storia
Naturale di Milano,
Milan, Italy, 2019.

Encounters with organic extinct animal remains—the most precious objects owned by many natural history museums since they are irreplaceable—are nearly always mediated by glass. Like Steph Berns discovered when studying encounters between museum visitors and religious objects in glass cases, glass can be distancing, but it can also facilitate encounters that would otherwise be unavailable to the public: "It is the case or nothing." The only things I ever encountered outside a sealed glass case or a glass frame were models, except for one taxidermied Japanese sea lion on display in the museum of the Osaka zoo. In some museums, like those in Vienna and Milan, their organic specimens of extinct birds are in fact behind two layers of glass,

one for the general display case and one for a specimen-specific box. This creates distance, but it also facilitates encounter by protecting the specimen.

Most natural history museums display one (sometimes a pair) of each species. In this way, a particular individual—the one behind the glass that the visitor is peering through—comes to stand in for all of its kind. There are instances in which the specimen's individual history is shared with the public, but this is rare. Instead, one passenger pigeon can stand in for all passenger pigeons. This has been the standard approach of natural history museums since the nineteenth century. It is because of this legacy that I write about *species*. But even in calling them species, we need to also recognize that the European Enlightenment project that brought about grouping animals into species is not the only legitimate knowledge system. Indigenous and local ways of knowing often stress other features such as specific habitat, particular coloring or marking, or even gender when assigning names to animal groups. Indigenous communities also do not necessarily think of extinction in the same way as Western science, often framing the disappearance of a species as the animal's withdrawal from relations with humans. Inspired by this emphasis on relationality, we can understand that extinction carries cultural and social meaning as much as scientific meaning.

Museums scholar Samuel Alberti has remarked that "the meanings of animals on display are imbued not only (if at all) by those with custody of them but also by their audience and therefore vary according to the ways they are looked at." It is in this vein that we can walk through the galleries of museums and investigate museum displays of the extinct. Rather than being concerned about what the creators intended to do with their display (authorial intent), we encounter the display on its own terms (reception). The "afterlives" of animals, as Alberti calls it, is a history of consumption as well as production.

There is a second glass in play in the extinction encounter: the camera lens. I documented my encounters, as most twenty-first-century museumgoers do, with my phone camera. Theopisti Stylianou-Lambert studied the use of cameras by art museum visitors and found six main motivations for taking photographs: to aid memory; to share the images with others who are not there; to enable further research/education after the visit; to inspire them after the visit; to strengthen self-identity; and to create their own art. These reasons apply in the natural history museum as well. Another study of visitors photographing with smartphones in the natural history museum in Gothenburg found that visitors both documented and shared their experiences. In the documentation processes, visitors made choices about what to photograph and how to photograph it, including image manipulation with smartphone settings. In the sharing processes, they posted content online and facilitated interaction beyond the museum walls. The study concluded that smartphone photography allows expanded interaction with items on display and may enrich the experience.

I certainly see my own photographic practice reflected in these descriptions, as I have taken photos both as documentation that aids my memory and enables further study, and for sharing with others. My gallery photos are my own artistic products and have often inspired my writing and thinking. My self-identity as a historian of extinction is wrapped up in my photos of the extinct.

I share my photos throughout this book to allow you to venture into the galleries alongside me. These are not professional photographs of natural history specimens taken in a studio with special lighting and equipment. The point of them is not to record a specimen scientifically or to be hung on the art museum wall. As museum visitors will notice from their own images, photos in galleries have shadows and reflections from light on the glass, and they sometimes capture nearby specimens and text. This is what the visitor experiences armed with a smartphone. These are everyday encounters, just like the thousands that occur daily in museums around the world. I have curated the photos—picked from among the thousands I have taken in the course of this research—so they don't represent everything I have seen, but they do offer a glimpse into the different displays. Each photo, as much as the text you are reading, invites you to stand with me in front of the glass peering at a ghost.

The animal remains and the context into which they are placed matter to the ghost's story. We will explore these configurations to expose how museums' presentations of the extinct affect our encounters with them. The goal is to help decode, contextualize, and analyze extinction displays to allow the extinct to tell their stories. Rather than going species by species or museum by museum, we thematically explore encounters with ghosts behind glass. Chapter 2 examines the ways in which the uncertainty of extinction plays out in choices made in museum displays. In chapter 3, we explore how extinct remains are framed as treasures; chapter 4 follows with how they are framed as cultural heritage. Then chapter 5 turns to the rare instances of named individuals on display when these individuals are considered the last of their kind. Next, chapter 6 discusses the broken relations of generations that often go unremarked in the museum. In chapter 7, we see the impulse to memorialize the dead with lists and collections. In contrast to the somber tone of memorialization, chapter 8 takes up playful interactions with the extinct. After venturing across the globe to visit all these museums, chapter 9 addresses extinctions the visitor typically will not encounter. In the epilogue, I reflect on why we should encounter extinction in the museum. Travel with me through the galleries of the world and encounter ghosts behind glass.

2 Seeing Ghosts

The striped body appears to walk through the shadows of the case at ground level. The face of the thylacine (*Thylacinus cynocephalus*), which was Australia's largest carnivorous marsupial, shines in the light as it comes into view. Its long snout seems ready to emerge from behind the glass, leading its slender, sleek striped body into the room. The back full of stripes makes it easy to see why Australian colonizers called it the Tasmanian tiger.

It stands in a dark room filled with glass cases with dark wooden frames. Each animal is spotlighted, fixed in time. They are attractive, like dramatic artworks in a gallery. But this thylacine I encountered in the Hall of Threatened and Extinct Animals at the Muséum national d'Histoire naturelle (MNHN) in Paris, France, attracts me not just as a ghostly figure of light and dark, but because of the historical narrative told by its label.

FIG. 15 Thylacine on display in the Muséum national d'Histoire naturelle, Paris, France, 2018.

Most of the specimens in this room are given short histories, explaining how or why that species is fit for this room of threatened and extinct animals. The labels typically give a geographical setting for the animal, a time frame for the extinction, and some indication of the reasons for its loss. The thylacine's story told when I first saw this gallery in 2015 went like this (my English translation of the French label):

> The species disappeared from the Australian continent because of the expansion of dingoes, and existed until 1960 on the island of Tasmania, where it was the largest carnivore. The thylacine ate all kinds of animals, notably kangaroos, wallabies, and ground-nesting birds. But after the colonization of Tasmania by the English, it abandoned kangaroos for sheep, which triggered a frantic hunt. Called the wolf of Tasmania, it got an overrated reputation as a monstrous nocturnal predator. For many years, bounties were offered for its destruction by authorities. In sixteen years, 2,268 thylacines were killed. Protection measures were passed late in 1936 by the Australian government. The last individual, a young male, was killed in 1961.

What instantly struck me about this story is that it is not the one I knew about the thylacine. The common story is that the thylacine became extinct in 1936.

Why was this story different? Was the museum seeing ghosts? Through the stories of the thylacine and ivory-billed woodpecker, this chapter explores how uncertainty enters into museum extinction stories. Although we might think that scientists would know if something were extinct, it is never that easy. Narratives diverge in the face of uncertainty, creating a plethora of stories, and we are never quite sure which to believe.

White settlers in Australia encountered thylacines, known col-
loquially as the Tasmanian wolf, tiger, or hyena depending on
who was writing, when they came to the island of Tasmania.
The thylacine, the largest marsupial carnivore in modern Aus-
tralasia, was systematically hunted by white colonial settlers
from the early 1800s because it was thought to predate on settler livestock. In 1888, the Tasmanian government established
a bounty on thylacines, and more than 2,100 payments were
made between 1888 and 1912. In addition to the government
bounty scheme, private bounties were offered by the Van Di-
emen's Land Company, among others.

From the mid-1800s, the thylacine appeared to be rapidly
disappearing (intentionally) from its natural range. In his *History of Tasmania* from 1852, John West predicted that "as every
available spot of land is now occupied, it is probable that in a
very few years this animal, so highly interesting to the zoologist,
will become extinct." In his well-known book on marsupials
from 1896, Richard Lydekker acknowledged that "a relentless
war of extermination" by settlers protecting their sheep "has
resulted in the almost complete extinction of this, the largest
of the Australasian Carnivores, in the more settled portions
of the country." Lydekker's implication was that although the
thylacine was not being seen by settlers, it was still numerous
in other, more remote areas.

Those hunted thylacines became highly prized collector
items, and many made their way into European museum collections. Tasmanian naturalists had already sent thylacines to
seven European collections by 1850. I have seen twenty-eight
museums that displayed some kind of thylacine remains. While

thylacines are obviously on show in the Australian museums, thylacines are also on display in museums in Austria, Belgium, France, Germany, Japan, Norway, Poland, Spain, Sweden, the United Kingdom, and the United States. Their bodies moved around the world as a curiosity: the largest marsupial predator that had the body shape of a dog and stripes of a tiger.

When the thylacine that was held by the Beaumaris zoo in Hobart, Tasmania, died suddenly on September 7, 1936, it was assumed that another thylacine could be found to replace it. In March 1937, the City Council of Hobart offered £40 to anyone who could bring in a live thylacine in good condition for the zoo. The zoo had to quickly retract the offer, because the Tasmanian Animals and Birds' Protection Board (TABPB, later to become the National Parks Service) had recently instituted a complete ban on thylacine hunting.

TABPB organized an expedition to find the elusive thylacine the next year. Tasmanian naturalist Michael Sharland led the expedition into the unpopulated mountainous regions where the animals were assumed to be living, away from human contact. The expedition made plaster of paris casts of thylacine footprints, but no thylacines themselves were spotted. This did not deter Sharland from believing the thylacine still survived: "It must be emphasised, however, that its failure to reveal itself more frequently is not necessarily indicative of approaching extinction. Great areas of this game country are devoid of human inhabitants, whilst others are only sparsely inhabited." Sharland believed that the animal was still present in remote Tasmania. The conclusion of Sharland's report was that a sanctuary should be declared to protect the remaining thylacines to avoid further encroachment on their territory.

In the 1940s and 1950s, there was still confidence that thylacines were in these remote areas. In 1949, the Taronga Zoological Park in Sydney received a permit from the TABPB to catch a pair of thylacines for conservation breeding purposes if they could be found, but the search was unsuccessful. An article in Melbourne's newspaper from June 1950 that asked "Can they be saved?" was a typical way of talking about the thylacine: it was "rarely seen" and "in danger of becoming extinct," but it was still around. When the Sydney Zoo applied again in 1954 for a permit to catch a thylacine, the permit was denied not because there were no more thylacines, but because the animals were considered too rare to warrant catching one. The belief that thylacines were still out there existed for decades, and searches continued to be sponsored.

But no thylacines were ever found, and eventually governments and scientific organizations decided the search was futile. The thylacine was officially declared extinct by IUCN in 1982 and by the Tasmanian government in 1986, although that hasn't stopped some people from believing that thylacines are still out there. Reports of (unconfirmed) thylacine sightings still appear in media even in the twenty-first century. There is a website that catalogs the extensive sightings of thylacines. The Tasmanian Department of Primary Industries, Parks, Water, and Environment regularly investigates thylacine sighting reports; there were eight investigated reports between September 2016 and September 2019, according to official records. In 2021, there was a spate of media coverage of a supposed thylacine filming, but scientists who viewed the images quickly dismissed it as a Tasmanian pademelon. The story of the thylacine's extinction is a hunt for ghosts.

The Tasmanian Museum and Art Gallery (TMAG) in Hobart has the largest thylacine display for the public anywhere in the world. When you walk into the room, you are greeted by three forms of thylacines: a taxidermy mount, a prepared skin, and two skeletons. Historical photographs of thylacines, particularly dead ones with hunters, fill one wall, while a film plays on loop on another: in black and white, the thylacine paces back and forth in its cage, a trapped animal trapped in time. This thylacine is the individual that died in Beaumaris zoo in

1936—all the moving images we have of a thylacine are this individual, whose ghostly image has to stand in for all thylacines. Still images as well as moving ones from this same film are often included in thylacine displays around the world.

The introduction text to the room explains that the thylacine is "a symbol of the island" as well as "a powerful reminder of how easily a species can be lost." The label on the wall adjacent to the film screening laments the thylacine's "sad history" that ends abruptly: "The last known thylacine died in the Hobart Zoo on 7 September 1936." In this declaration, TMAG is following the generally accepted extinction story: the one that died in 1936 was the last, and all sightings after that are ghost sightings. There has been a rhetorical convergence on September 7, 1936, as the extinction of the thylacine, despite reports of animals alive in the wildlands of Tasmania long after that date.

At the same time, the TMAG exhibition acknowledges the continued search for thylacines. A case on the same wall as the film holds physical artifacts of the searches: a plaster cast of the footprints collected by the Sherland expedition; a "Thylacine response kit" issued by the Tasmanian Parks and Wildlife Service in 1983 to their parks officers to ensure that evidence of thylacines would be properly collected; a remote sensing camera that was set up after a reported sighting; and reward posters for those able to turn in definitive evidence of thylacines still alive. But the narrative is still that all these searches came up empty handed. The thylacine that died in Hobart in 1936 was the last.

Although nearly all displays of thylacines (and there are many such exhibits, because thylacines are one of the most common extinct species held by museums around the world— see the appendix), like the TMAG exhibition, give 1936 as the

FIG. 16 (*facing*) Two Thylacine skeletons seemingly watching the moving image of the Hobart zoo's last thylacine in the Tasmanian Museum and Art Gallery, Hobart, Australia, 2016.

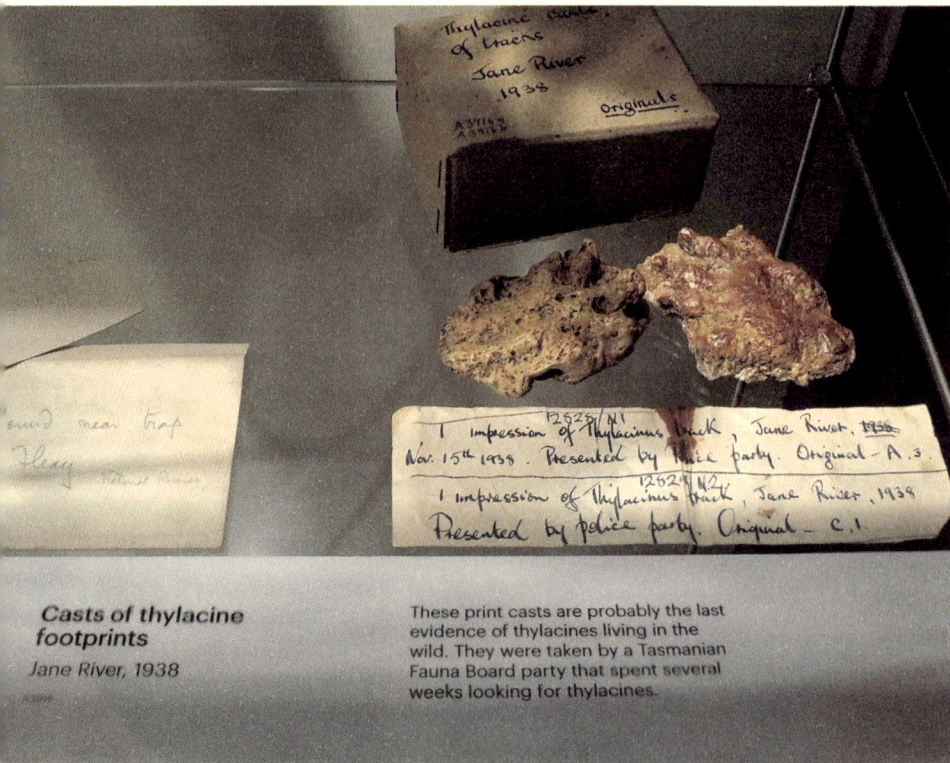

Casts of thylacine footprints

Jane River, 1938

These print casts are probably the last evidence of thylacines living in the wild. They were taken by a Tasmanian Fauna Board party that spent several weeks looking for thylacines.

FIG. 17
Display of evidence from thylacine searches, Tasmanian Museum and Art Gallery, Hobart, Australia, 2018.

extinction date, alternative histories of the last wild thylacine and later sightings make appearances. While many official narratives have seemingly coalesced on the death of the Hobart thylacine (erroneously called Benjamin in some displays) as the species extinction point, other museums display diversity in their historical narratives.

In the Museum für Naturkunde (MfN), Berlin, for example, the label for the thylacine says (in my English translation), "In 1930 the last free-living thylacine was shot." The last documented wild thylacine was an adult male shot by Wilf Batty on May 13, 1930, on his farm. A picture of a grinning Batty posing with his shotgun, dog, and dead thylacine features in the TMAG exhibit and in books on the thylacine. By choosing to narrate the extinction as the death of the last confirmed wild animal, rather than one that was confined in a zoo, MfN adopts an ethical position on the relation between free living and worthwhile living. Although zoos have been key players in the protection and recovery of some endangered species, including ones that were extinct in the wild, like the European bison and California condor, they have also been the final home of the last known individual of a species, including the thylacine, the passenger pigeon, and the Carolina parakeet. In MfN's choice to tell the story of the last wild animal rather than the last captive one, the role of the zoo in extinction is elided.

In their *Extinction Voices* temporary intervention, the Bristol Museum and Art Gallery decided to stress the death of the individual thylacine in their own display, rather than the species as a whole. Their specimen was given a life history through the signage: "On 28th December 1914 London Zoo telegrammed Bristol Museum. 'Will you take Tasmanian Wolf, five pounds, almost extinct.' This male thylacine had died on Christmas day. It was one of a pair bought by London Zoo from Tasmania in 1910. The female hadn't survived the journey." The text was accompanied with a photograph of the thylacine in the London Zoo when it was still alive. An adjacent sign noted that "Twenty-one years after this specimen arrived at Bristol Museum, thylacines went extinct. When Bristolians first saw it on display, thylacines were still alive in the world. Now you can only see one in a museum." In this extinction narrative, the thylacine's eventual extinction is just one part of the story—the museum's acquisition of a specimen of the species, which was recognized as extremely rare at the time, is the main message.

The extinction story summarized in the MNHN display in Paris that I encountered in 2015 (and which has since changed) was another story altogether. Here, the museum told a story of a last thylacine in 1961. This wasn't a coincidental date. In 1962, Sharland, who had led the 1938 expedition to hunt for the thylacine, published a book titled *Tasmanian Wild Life*. In that book, he tells the story of two fishermen in 1961 who claim to have encountered and then killed a young male thylacine at night without realizing what it was. They supposedly put the corpse into their fishing hut, returned to fishing, and then when they returned, the body was missing. According to Sharland, "Apparently in their absence some person had entered the hut and stolen it! The men were upset about the disappearance of so valuable a piece of evidence and reported the matter as soon as they got home." They sent some blood smears and hairs to the university for confirmation, which Sharland claims proved "both the hair and blood referred to a Tiger and nothing else." Later writers would claim that the blood was too degraded to test and the hair was not definitively a thylacine; supposedly one of the fisherman, Bill Morrison, changed his story in 1980 to say it was a Tasmanian devil rather than a thylacine. In spite of these later doubts, Sherland's claim was the one used for the MNHN display, which was opened in 1994, not long after the thylacine was declared extinct. The curator of the exhibition, Florence Raulin-Cerceau, had written the label for the thylacine. She repeated the story that "the last individual, a young male, was killed in 1961" in an article published in 1994, so it clearly was an intentional narrative.

There is even an example of the thylacine being displayed as not extinct at all. The Naturhistorisk museum of the University of Oslo, Norway, has a thylacine standing in a green case under a map of the world. The map is interactive: the visitor can press a button for five different animals and lights appear that show that animal's distribution in 1870, indicated by red lights, compared to the animal's distribution in 2009 with green lights. When I pressed the thylacine light in 2014 during my visit, the result was that a green light appeared in Tasmania. The label under the specimen asks, "Is it extinct?" and the light appears to tell the visitor that the answer is "no." The doubts raised by continued sightings had clearly influenced the exhibition content. Although the thylacine had been declared extinct by IUCN over seventeen years before this exhibition opened, it is portrayed as potentially still alive.

In other displays, this doubt of extinction seeps in in small

FIG. 18
Thylacine with its habitat of Tasmania indicated with a green light, Naturhistorisk museum, Oslo, Norway, 2014.

FIG. 19 The thylacine under doubt (*twijfel*), Gents Universiteitsmuseum, Gent, Belgium, 2022.

ways. At the Smithsonian Museum of Natural History in Washington, DC, the thylacine's label has a large red stamp with "presumed extinct" next to the text that describes the hunting traits of the "extinct thylacine." While the main text uses "extinct" without a modifier, the more prominent stamp does not. Doubt is raised about its extinction status.

This is not a case of bad science. The thylacine displays at MHNH and the Naturhistorisk museum reveal the contestation inherent in declaring a species extinct. When a species is declared extinct, it is believed that it no longer exists anywhere. But having scientific proof of such nonexistence is impossible because scientists cannot see all places simultaneously. There's always a possibility that one individual is alive in a place not observed at that moment. This is a problem of converting the absence of presence (a failure to find the animal in the places that are searched) into a presence of absence (it exists nowhere). To say that a species is extinct requires moving from one to the other. Because it is physically impossible to see everywhere at once, declaring a species extinct is tricky. There can, of course, be good scientific reasons for assuming a species is extinct, but there is never 100 percent certainty. All extinction stories are haunted by the possibility that the extinct may still be alive.

The thylacine display at Gents Universiteitsmuseum in Belgium speaks to this uncertainty. The museum has organized its collection through keywords rather than by taxonomy or geography—and the thylacine stands proudly under the sign "Twijfel" / "Doubt." The doubt in this display is not a doubt of the thylacine's death, but rather its life: the label discusses the use of specimens like this one to "reconstruct how this extinct animal used to live," yet there is always "a significant risk of

misinterpretation." Based on encounters with thylacines in museums around the world, doubt indeed surrounds the thylacine.

HAUNTINGS IN THE TREETOPS

The ivory-billed woodpecker (*Campephilus principalis*) has
been haunted by similar possibilities. The ivory-billed woodpecker was the largest woodpecker species in the United States
at the time it was colonized by Europeans. Clearing of the
southeastern bottomland hardwood forest caused major declines of the woodpecker population in the 1800s, and then, because of their increasing rarity, the birds were favorable targets
for collectors (both lay and scientific) who wanted to ensure
that the bird would be represented in their private or public
collections. The birds and their habitats were ravaged by re-
source extraction. By the time ecologist James Tanner inten-
sively studied the birds in the late 1930s, there were probably
two dozen ivory-billed woodpeckers left. The last universally
accepted sighting was in 1944 on the same tract that Tanner
had worked in.

Yet there was always a lingering doubt about the woodpecker's extinction. Reports of the bird trickled in from the 1950s to
the 1970s. In 1967, the bird was listed as endangered. Evidence
of the bird's continued presence was always tenuous. For example, audiospectrogram analysis of a 1971 tape recording of a
bird call couldn't determine whether the call was an ivory-billed
woodpecker or a blue jay. In an exciting turn of events, in 2005,
scientists associated with the well-respected Cornell University Laboratory of Ornithology announced that they had audio
and visual evidence of ivory-billed woodpeckers in Arkansas,
including one in flight, from 2004. The announcement caused

a media frenzy, but the visual evidence was quickly refuted by other experts. Numerous extensive searches were made to look for more evidence, but nothing incontrovertible was found. Geographer Matthew Gandy has remarked that "tired ornithologists in their waders and canoes, batting away mosquitoes, and hoping against hope to make a famous sighting" were "haunted by the absence of the ivory-billed woodpecker." The continued presence of the ivory-billed woodpecker as a specter motivated the continued desire to find it. The Laboratory gave up the search after a few seasons. Yet the US Fish and Wildlife Service (USFWS) continued to consider the bird as endangered and even issued a new species recovery plan for it in 2010.

The liminal status of the ivory-billed woodpecker as *possibly* or *probably* extinct, rather than just extinct, is visible in many museum displays of the bird: it is "probably extinct by 1944" in the Smithsonian Natural History Museum; in Manchester, it is "extinct?" with a question mark; and in the Harvard Museum of Natural History and the Royal Albert Memorial Museum and Art Gallery in Exeter, the label is "Critically Endangered." The Naturhistorische Museum Wien announces that while the ivory-billed woodpecker is extinct from the southern United States, it is still found in Cuba. While there is indeed a Cuban ivory-billed woodpecker, there is scientific debate about whether it is a subspecies of the Continental one or its own species. In any case, Vienna is marking the species as still extant. In the Naturhistoriska riksmuseet in Stockholm, the label also connects the birds in Cuba and the southern United States, but notes that the last Cuban one was seen in 1987 and in 1996 it was listed as extinct. Then the label goes on to recount the Cornell Ornithology Laboratory sighting: "In 2005

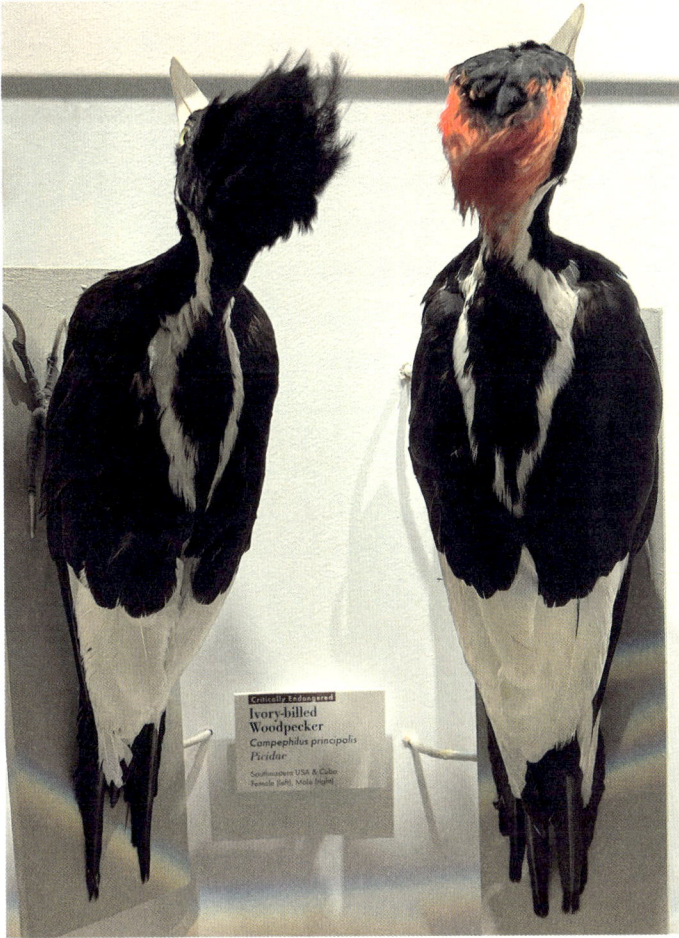

FIG. 21
A pair of ivory-billed
woodpeckers categorized
as "critically endangered"
at the Harvard Museum
of Natural History,
Cambridge, Massachu-
setts, USA, 2021.

it was claimed that the woodpecker had been rediscovered and
filmed deep in the Arkansas swamps in the USA." While the
bird is placed clearly in an "extinction" case, the label portrays
it as still potentially living.

In the California Academy of Sciences, a pair of ivory-billed woodpecker models hold tight to a giant cypress tree model. They are part of the scenery for the Cypress Swamps exhibit, which features the famous albino alligator Claude. The large information board asks, "Lost or found?" The first statement is that the bird "was believed to be extinct." The label then offers up the 2004 "multiple individual sightings and a brief blurry video" as evidence of presence and states that scientists "have been working to confirm if the species still exists."

Even when the display does note the ivory-billed woodpecker as extinct, there are inconsistent histories of extinction. The MNHN label says that the last certain observation was in 1962—although this is not a date generally agreed upon in the scientific literature. Where did this date come from? In bird enthusiast Tim Gallagher's book *The Grail Bird*, he discusses a search in 1962 on the same track where Tanner had seen the last confirmed bird. During the search, four graduate students heard two bird calls that sounded like the 1935 recording of ivory-billed woodpeckers, but identified the birds as a northern flicker and a blue jay. The students were sure that those birds must have been mimicking an ivory-billed woodpecker call, so they continued the search. They heard another bird with the right call and one of them caught a glimpse of it, identifying it as a large woodpecker, and were sure that it was an ivory-billed woodpecker. This 1962 sighting must be the one that MNHN is referencing on the label.

In September 2021, seventy-seven years after the last definitive sighting in 1944, the USFWS proposed removing the ivory-billed woodpecker from the endangered species list because it should be classified as extinct. The proposal received 157 comments in the regulatory portal, nearly all of which were asking that the bird not be delisted. These commentators included biologists, hunters, and bird enthusiasts. Shane McCorristine and William Adams have noted that "absence exerts a powerful influence among conservationists as they, more than most scientific practitioners, patrol the ghostly boundaries between what is here, going, and gone." The absence of the woodpecker as a sign of endangerment is actionable, whereas absence as a sign of extinction is a dead end. In the wake of the flood of comments, the USFWS decided in 2023 to *not* delist the bird, which means it is still officially considered endangered in the United States.

The hesitation to list a species as extinct is due to uncertainty, and, in fact, species can be rediscovered. In the Grant Museum of Zoology, the labels on two Australian Central Rock rats (*Zyzomys pedunculatus*) indicate that the species has been declared extinct twice, but is now considered extant, although critically endangered. This kind of rediscovery history is not a lone occurrence. The Australian night parrot (*Pezoporus occidentalis*) had no confirmed sightings between 1912 and 1979, but since 2016, several live night parrots have been filmed and a new colony identified in Western Australia, and the Lord Howe Island stick insect (*Dryococelus australis*), which was believed to have been extinct since 1930, was rediscovered in 2001. Such a discovery of a long-lost species is "the ultimate prize of conservation ghost-hunts," according to McCorristine and Adams. The ivory-billed woodpecker continues to act as a ghost, a prize to potentially be won if a sighting can be confirmed.

What happens to the ivory-billed woodpeckers on display in museums if the species' status changes? Will the label in the Harvard Museum of Natural History be updated to "Extinct"? Will the question mark be taken off the one in Manchester? As long as the ghost hunt continues, it is difficult to come to definitive answers—to claim, for once and for all, that this woodpecker is extinct.

UNCERTAIN HISTORIES

Encounters with thylacines and ivory-billed woodpeckers in museums are always encounters with history. Natural history museums typically don't tell modern histories. They might tell evolutionary histories of relations between species or convergent evolutionary histories of traits (such as the development of wings), but they generally don't give a species a history in recorded time. The general exception to this is endangered and extinct species, which are often displayed with tales of historical decline in time and space. The historical narrative of the extinctions of these species is told with a label like "Critically Endangered" or through a longer text giving dates and events leading up to the species' demise. Yet because extinction histories are so often uncertain, a single historical narrative is elusive. These histories are anything but simple.

Because extinction is uncertain and new information on extinct species is always becoming available through new scientific studies, museums and their displays are not static. When I revisited the Paris museum in 2022, the thylacine had been rehoused in a new case with a new label. Now, the label read "The last wild individual was captured in Tasmania in 1933 and died in a zoo in 1936." The 1961 story had been thrown out in favor of the more orthodox extinction story. Here, a museum label was coming into line with the most common historical narrative. With the ivory-billed woodpecker, the histories are on even less sure footing. While the last universally confirmed sighting in 1944 is recounted on the Smithsonian label, it is the only one that I found tells this history. While the bird is marked with a symbol as extinct in Toronto's Royal Ontario Museum, most museums adopt a more cautious tone about the bird's extinction. "This bird is critically endangered and may even be extinct," the label in Exeter proclaims. The status is not clearcut because there are still some claims that it is alive.

The uncertainty of extinction—the continuing ghostly presence of the extinct—haunts the stories of extinction told through museum labels and signage. The ongoing searches for thylacines and ivory-billed woodpeckers cast a shadow over museum displays. Because most visitors to a natural history museum do not visit museums across the world, they will not see the variety of stories; instead, they will see the extinction story at one museum and assume it is the one and only story. Yet, as these thylacine and woodpecker encounters show, the histories are multiple and divergent because of uncertainty. Ghost stories are difficult to pin down.

3

Cursed Treasures

Visitors walked under the words "Treasure chamber"—the *Schatkamer*—into the box in the middle of the room. Their path had already passed by one display of specimens, including an extinct great auk, on the way to the doorway. The gold-colored tiled walls reinforced the words above the doorway, calling attention to the fact that this is a special place: a place filled with treasures.

The lighting was low, but the Cape lion shone out of the darkness. Looking around this treasure box, we could find the remains of three other extinct species: an elephant bird egg, a dodo skeleton, and a taxidermied quagga. Not everything in the room was extinct—there were also objects like a sixteenth-century herbarium and an opossum that was a specimen described by Linnaeus himself. What the objects all had in common is that they had all been selected by Naturalis, the newest incarnation of the Dutch National Museum of Natural History, as part of the special exhibition *Nature's Treasure*

Trove in Leiden, the Netherlands, to commemorate 200 years of collecting.

This was not the first time that I had been in a treasures gallery to encounter the extinct, nor would it be the last. I had previously seen the permanent exhibition, *Treasures*, in an upper-floor gallery of the Natural History Museum in London. The long, skinny room is like a jewel box, replete with stained glass windows on both sides. There is definitely a spiritual air about the room, which in some way contains the holy of holies, at least as far as natural history specimens go. Twenty-two objects are in the room, including many linked to famous naturalists like Alfred Russel Wallace and Sir Hans Sloane. The great auk and a composite skeleton of the dodo are included in these treasures.

The likeness of the *Treasures* gallery to a cathedral with stained glass windows is not arbitrary. When the Natural History Museum in London opened in 1881, an article in *The Times* called it a "true Temple of Nature" that demonstrated "the Beauty of Holiness." The building is in the twelfth-century Romanesque style used in cathedrals and churches, which the architect Alfred Waterhouse and the museum's curator of natural history Sir Richard Owen felt communicated both grandeur and simplicity. The building captured the epistemological and theological motives of Owen: that science and the material reality of nature conformed to "sacred, metaphysical principles." The religious aspects of the building's design were intentional. Natural theology was built into this and other Victorian natural history museums.

A couple of months after my visit in Leiden, I visited another treasures gallery, the *Precious Things* exhibition at the Statens Naturhistoriske Museum in Copenhagen. This gallery

FIG. 23 (*facing*) The explicitly religious design of the *Treasures* gallery at the Natural History Museum, London, UK, 2022.

of items picked out by the museum staff as the highlights of the collection expands like a dark labyrinth before the visitor. The lights are low, and there is no single path to follow through the exhibition. It's easy to get lost here. You are never quite sure if you have taken every path and seen all the displays. It gives the visit the feeling of a treasure hunt: What amazing thing will you see around the next corner? Some of those amazing things include the extinct: thylacine skeletal remains, a great auk collection, and a group of giant anole lizards in jars. The most precious of these gets a vault of its own: a unique complete dodo skull sits inside a sealed room with a thick door with no handle and an electronic keypad to disarm an alarm.

Being extinct assigns value. An example is a lion that was once a royal treasure, then trash, then treasure again. This Barbary lion lived in the 1400s in the royal menagerie at the Tower of London, an animal worthy of the king. But the animal was only valuable when alive—the skull was excavated in 1937 from the tower's moat, where it had been thrown out as worthless waste after the lion died. The skull is now on display in the Natural History Museum's treasure room because north African Barbary lions are extinct.

What kind of treasures are the remains of the extinct? A walk through treasures exhibitions allows us to explore the issues of rarity and irreplaceability with extinction. We focus in on two species that once inhabited the area now known as South Africa: the Cape lion and the bluebuck. When museums put organic remains of extinct animals on display, they are showing an item that can never be replaced. Whereas a model can be copied and displayed in many museums across the world, the number of skin and bone specimens of an extinct animal are limited, so having one is special.

There are three mounted adult Cape lion specimens on display in the world. Two appeared to shimmer in front of their black surroundings: the one in Leiden stood among the rare treasures, and the one in Paris stood among fellow extinct and endangered animals. The exhibition design of both displays highlights the specimens: while there is a label of white text on black next to the animal, there is no patterned background, pictures, or extraneous information to distract the gaze. The focus is on the animal, as perhaps it should be with such a rare thing. The third lion is displayed in a completely different context: it is at Clifton Park Museum in Rotherham, UK, within a children's activity room. We will encounter this particular Cape lion in chapter 7 on playing with ghosts.

The Cape lion (*Panthera leo melanochaita*) was a distinct subspecies of lion largely confined to the region of the Cape Province in South Africa. The Cape lion was first named in Western scientific literature as *Leo melanochaitus* in 1842 by Charles Hamilton Smith, a former British military officer and well-published naturalist. In Smith's description and the illustration accompanying it, the most remarkable characteristic of the animal was its black mane and a fringe of black hair under the belly. Smith had earlier sent the drawing to Edward Griffith for reproduction in *General and Particular Descriptions of Vertebrated Animals* in 1821. The "Black Maned Lion of the Cape" was considered its own species at the time, although recent research has demoted it to a subspecies of African lion (*Panthera leo*).

The arrival of Dutch then British colonial armies from the mid-seventeenth century put direct pressure on the Cape lion population. The lions were collected for local and homeland menageries and hunted as trophies. For example, the particular lion illustrated by Smith had been "taken at the Cape, and was presented to Lady Castlereagh." Lady Castlereagh (Amelia Stewart) had a private menagerie at her country home Woollett Hall near London, so this lion was shipped to live in her private zoo. Cape lions were captured and sent to other English collections as well, including the Tower of London menagerie. Many more were killed as trophies, for museum displays, or as agricultural pest control. Although lions had been commonplace near Cape Town at the beginning of the 1800s, by the 1860s, they were extinct.

In 1962, lion specialist Vratislav Mazak did a survey of museum holdings and found six mounted specimens of Cape lions: a male in the Natural History Museum in London, a male in Leiden (now Naturalis), one male and one female in Staatliches Museum für Naturkunde in Stuttgart, a male at Städtisches Museum Wiesbaden, and a male in Musée Nationale d'Historie Naturelle in Paris. Most of these are not on display. The Cape lion displayed as a Dutch treasure had made its way into the Leiden collection in 1860 when it was transferred from the Cabinet d'Anatomie in Utrecht. Where and when it had been acquired by that collection is unknown. The Cape lion in Paris was a lion caught at the Cape of Good Hope (Cap de Bonne-Espérance) that died in the menagerie in Paris in 1834. The French artist Antoine-Louis Barye may have drawn this lion from life for a watercolor that was exhibited for the first time in 1833. There is one photograph from 1860 of a living Cape lion that was also in the menagerie of Jardin des Plantes, but that is not this individual.

In the Dutch exhibition, the rarity of Cape lions was not

FIG. 24
The Cape lion (*left*)
and Barbary lion
(*right*) on display at
the Muséum national
d'Histoire naturelle in
Paris, 2022.

highlighted; rather, the Cape lion's ubiquity was stressed. The
label contains a paradox: "The last example of a Cape lion in
the wild was observed in 1865. In the Netherlands we can still
see the Cape lion: three of them feature on the Royal Coat of
Arms." Laura Bertens and Ann Marie Wilson have remarked
that this "sudden and somewhat jarring pivot away from its

account of the lion's extinction" is a way of glossing over the implications of colonialism in the exhibition. The Cape lion is a treasure in *Nature's Treasure Trove* not just because it is rare, but also because it evokes Dutch national identity. The Cape lion as symbol has in fact become more common than the animal ever was in South Africa because of its appearance in the Dutch coat of arms. Scholar Audra Mitchell has noted this tendency of the extinct, while absent, to proliferate culturally. That's what the Cape lion has done—it has been reproduced exponentially despite being unable to reproduce.

This is similar to Tasmanian tiger images, which are everywhere in Tasmania. There are thylacines peeking out of long grass on vehicle license plates, standing on the coat of arms for Tasmania (just like the Cape lions on the Netherlands' one), and on the old logo of Cascade beers. Thylacines are on tea towels and t-shirts for sale in gift shops. It is the same with the dodo on Mauritius. Dodos proliferate on cheap fridge magnets and high-end jewelry sold to tourists from the shops that spill over into the street. Mauritius is the land of the dodo, with the dodo often appearing alongside outline maps of the island nation on souvenirs. But dodos are not just a gimmick—they also appear as the watermark on the Mauritius rupee bank notes and as the image on the official immigration registration card. These animals are specters that continue to move through culture. Perhaps through their appearances, they distract us from their rarity and the finality of their extinction.

IRREPLACEABLE REMAINS

More than being just rare, the Cape lion remains held in museums are irreplaceable. There is no possibility to acquire a new one from the wild, and the sale or exchange of extinct animal remains is exceedingly unusual. This means that those museums holding the remains of the extinct have to maintain them or risk losing them forever. All organic materials—whether those are skins or organs or skeletons—are constantly in the process of degradation. To maintain these remains is to fight against entropy. The struggle to keep them from splitting apart at the seams is real.

We see this when we peer through the glass at the bluebuck (*Hippotragus leucophaeus*) on display in Naturhistoriska riksmuseet in Stockholm. The Cape lion was not the only animal endemic to the Cape that became extinct in the area after European colonization; others include the blaauwbok / bluebuck / blue antelope, quagga (*Equus quagga quagga*), Cape warthog (*Phacochoerus aethiopicus aethiopicus*), and Cape hartebeest (*Alcelaphus buselaphus caama*). The bluebuck is even rarer than the Cape lion: museums hold only six specimens of any type (bone, horn, or skin) confirmed to be bluebucks, and another four are possibly bluebucks but unconfirmed. Only four mounted bluebucks exist anywhere, and only two are currently on public display within museums: one at the NMHM in Paris and the other at Naturhistoriska riksmuseet in Stockholm. The bluebuck label in the NMHM says there are five mounted specimens in museums, but this is incorrect. There appears to have been a bluebuck mounted specimen at some time in the past in Uppsala—many older publications reference it—but either the specimen has been lost/destroyed, or it was an error that took on a life of its own in the literature.

In Stockholm, the bluebuck's head appears perilously attached to the long neck, an opening seam revealing the stitches that had once created the illusion of wholeness. Instead of flesh, stuffing attempts to jut out of the cut. The specimen has obviously seen better days. But it is almost all there is. Without it and its counterpart in Paris on display, the bluebuck would disappear into the museum back rooms—out of sight, out of mind. But keeping its ghostly presence standing in the case, with its body ripping itself apart at the seams, is a reminder of the violence of extinction.

The Stockholm exemplar had been transferred to the museum in 1829 from the massive collection of industrialist Adolf Ulric Grill. The provenance of the animal is uncertain, but it is possible that Grill acquired it from fellow Swede Anders Sparrman, a naturalist who journeyed in the Cape region from 1772 to 1776. Sparrman describes finding a bluebuck and preserving the skin, which he says becomes more gray than blue after death. It is unclear from the sentence construction whether Sparrman saw the animal alive or if he just saw the skin. Scholars have tended toward the latter interpretation. Sparrman gave several other specimens from his Cape journey, including a Cape buffalo and a blesbok, to the Swedish Academy of Sciences, so it is possible that he gave or sold the bluebuck to Grill. Although Grill was most renowned as a bird collector, when Grill's widow Anna Johanna expanded the family nature cabinet in 1808, the mammals collection, including the famed bluebuck, took center stage. The Grill collection was transferred to the Naturhistoriska riksmuseet in 1829. When the history of the museum was written in 1916, the bluebuck was considered "one of the National Museum's greatest treasures."

The bluebuck, unlike the Cape lion, does not proliferate. It was so rare and disappeared so quickly after European colonization of southern Africa that it made no significant cultural mark. Bluebucks appear to have been restricted to the valley

of Soete Melk in Swellendam. The well-known naturalist and ornithologist François Levaillant and his hunting party success-fully killed and skinned one bluebuck in the wild in the valley in 1782. He noted that the animal was extremely rare; he saw only two during his five years in Africa, and both had come from the same valley. German physician and explorer Hin-rich (Henry) Lichtenstein, who traveled in South Africa from 1803 to 1806, remarked that the "almost extirpated" bluebuck was occasionally found in interior valleys of the Cape, yet they

FIG. 26
A full view of the
bluebuck in the
Muséum national
d'Histoire naturelle,
Paris, France, 2022.

had not been seen alive since 1800, when some were shot and
their skins taken to Leiden. According to genetic research, the
bluebuck population was already very small (estimated at 370
individuals) when the colonists arrived. Although the small
population may have been headed toward extinction anyway,
hunting by European colonists, including these rare specimens

demanded by collectors, must have hit the small population of bluebuck hard.

While the Stockholm display label is vague about the extinction—"the last bluebucks were killed in South Africa in 1800"—the Parisian display is forthright about the colonial context of the bluebuck's loss: the bluebuck "was one of the first African mammal victims of the colonization of southern Africa." The label in Paris then goes on to describe the bluebuck as "a small fearful animal, it was easy to hunt for meat, usually intended to feed dogs." This particular description is at odds with the large antelope on display. In missionary Reverend Latrobe's journal of his journey in South Africa in 1815–1816, he described the bluebuck as "remarkably fine; brown, changing with a blueish grey" and "not as swift as others of the same kind," so it could be caught by an Indigenous man. This must be the reference for describing the bluebuck in Paris as slow. But attributing these characteristics to the bluebuck is wrong on two counts: Latrobe had visited South Africa after the last bluebuck had died, and the description is for an animal called "one of the smallest antelopes," which would better match a blue duiker (*Philantomba monticola*) than a bluebuck. The confusion shows how little we know of the bluebuck.

The confusion continues to mount as time goes on, because we have even fewer bluebucks than we once thought. The Hunterian Museum, part of the University of Glasgow, has a skull with horns on display in a corner. The specimen is labeled as a bluebuck, and it is part of a display about extinction—a thylacine is in a case nearby. The label proclaims that "today only seven examples of bluebuck remains are known in the world." The problem is that this is not one of them: tests of mitochondrial DNA published in 2021 show that this is a sable antelope

(*H. niger*). The label and display had not been updated when I visited in 2022.

The bluebuck or blue antelope was only described by Western science in 1766 and was extirpated by around 1800. The earliest known full-body image of a bluebuck, drawn by J. Allamand and published in 1778 in *Histoire naturelle, générale et particulière,* by G. L. Leclerc, Comte de Buffon, was based not on a live bluebuck, but on the mounted Naturalis museum specimen (which is unfortunately not on display). The specimen used for the first bluebuck drawing was one of the very few bodies that survived the rapid extinction. Many natural history drawings of the time were not drawn from animals in original habitats—artists often worked from animals in menageries, from mounted specimens or skins, or even from oral or written descriptions without having seen the animal in question. This could produce problematic results if neither the artist nor the taxidermist had seen the animal alive. Levaillant felt this had happened with the bluebuck, writing, "I do not recognize this animal very much in the drawings and engravings that I have seen of it up to now," so he made a more accurate drawing of the bluebuck that was shot by his party before it was skinned. The drawing survives in two copies, one in the Library of the South African Parliament and the other in the University of Leiden Library. The individual drawn by Levaillant also survived—it is the animal on display in Paris.

This treasure, like most taxidermy, is just skin on a shell. Levaillant had described the quickness of the hunt—the Indigenous hunter "aiming at the animal and shooting it was a matter of an instant" and afterward "proceeded to skin the animal with the same skill as he had shot it." Yet cut marks on the skin of the Paris specimen show that it was not as meticulously prepared for taxidermy as other antelope specimens of the time. Presumably the local hunter who assisted Levaillant would have been more familiar with cutting up an animal for meat rather than for taxidermy purposes. The hasty demise of the bluebuck in the context of colonialization can be contrasted with the slow decay of the specimens. Looking at these bluebucks, we see the dead body fighting to break down to dust. The skin has contracted on the frame, making the seams split.

The bluebuck specimens on display all show significant wear and age, but this isn't necessarily a reason to remove them from the public eye. I encountered a Carolina parakeet (*Conuropsis carolinensis*) in the Zoological Museum of KU Leuven in Belgium long after I had seen these bluebucks. It sat perched on a stick formed into a stand that was placed on the floor of the vitrine. From my vantage point above it, the wispy colored feathers, seemingly barely attached to the skin, were visible. It wasn't the sleek parakeet on display in other museums; it was tattered and torn, wearing its age. At first sight, it was not pleasant to look at—its body seemed to reveal a violence that lurked behind it. But because it was so close to the ground, the tiled floor was reflected in the glass between the viewer and the parakeet, creating a view that might be compare to an impressionist painting. The feathers in this view were not just taxidermy in need of repair, but rather an aesthetic drawing. Looking at the parakeet from this vantage point, its purpose became more clear: that specimen needed to be there, not in spite of its condition, but because of it. The Carolina parakeet's condition is extinction, so why shouldn't it look disheveled and aged? Like the bluebucks on display, this parakeet's condition might actually facilitate reflection on extinction and death.

The few specimens of the extinct bluebuck that survive are

irreplaceable treasures. Because one of the mandates of the natural history museum is to preserve specimens for future study, it makes sense that some of the bluebuck remains are locked away. Yet the two individuals on display are the only reminders of its former life and extinction that the public visiting a museum will see. In seeing their skin stretch apart, we get a glimpse into that violent history.

THE TREASURE'S CURSE

There is a common folklore tradition that hoarded treasures are often cursed. From the dwarf's treasured jewel that makes its owner, whether dwarf or dragon, go mad in *The Hobbit* to the treasures of King Tut's tomb, which supposedly led to uncanny deaths among those who had excavated it, treasures might be dangerous. This danger comes in the form of a haunting by previous owners through a curse. Often the haunted treasure appears because of its acquisition—that it was looted or plundered to enter unworthy hands. The curse is typically only lifted when the treasure (or at least part of it) is returned to its rightful owner or final resting place.

Might we think of the extinct Cape lion and bluebuck treasures on display in the museum as cursed? They too were acquired in unsavory circumstances through rapacious hunting by white settlers. Their lives—all of their lives—were stolen. When we look at them on display, perhaps their glass eyes are cursing us for what humans in the past did to them.

But lifting the curse is not easy. There have been calls for returning cultural artifacts and human remains from museums that acquired them through colonial powers, and this is happening in many European and North American museums.

FIG. 27
Jamaican giant
galliwasp, which was
recently returned to
Jamaica, the Hunterian
Museum, Glasgow,
UK, 2022.

While the International Council of Museums Code of Ethics
for Natural History Museums includes a discussion of repatriation of human remains, repatriation of animal specimens
taken away from their original home is not presented as an
option. Yet there have been a few animal specimens returned

or requested. An endangered leatherback turtle acquired by the Te Papa museum in New Zealand was returned to the local Māori hapu for burial as a guiding deity in 2019, and Tanzanian politicians have asked for the return of dinosaur bones taken by German colonial powers and now displayed in the Museum für Naturkunde in Berlin.

One modern extinction specimen has recently joined these returns. In June 2024, a Jamaican giant galliwasp (*Celestus occiduus*), a lizard endemic to Jamaica that was last seen alive in the 1800s and is believed extinct, was repatriated. The Hunterian Museum in Glasgow turned over its specimen to the Institute of Jamaica for relocation to the Natural History Museum of Jamaica. I had seen this specimen on display in a section called "Curating Discomfort" about power imbalances within museum practices. The giant galliwasp, the largest member of its genus in Jamaica, had been an easy target for herpetologist collectors. But its extinction was likely due to introduced mongooses. These predators had been introduced to the island to combat rats on colonial sugar plantations, but it turned out that they also hunted the lizards. It is fitting that a specimen put on display in Europe to make colonial extraction visible is the first to be returned to its former colonized home.

But this kind of call for specimen return is not a widespread phenomenon. For now, the remains of extinct animals most often haunt the halls of museums far from home. Bodies of animals have been understood as noncultural, yet of course they are displayed in a museum precisely because they fit cultural concepts of science and what should be in museums. They are cultural artifacts no less than a bronze mask. And yet, we can ask, whose cultural artifacts are they? While these animals originally ran, leapt, grazed, or hunted in South Africa, they typically turned into artifacts at the hands of European hunters, taxidermists, and curators, although we must not discount the Indigenous experts who also facilitated and participated in the hunts. The movement of the bodies of extinct specimens away from their origin has created an "extinction echo" in which the animals are both lost physically long ago and then lost later in the local public imagination. Perhaps no humans where they once roamed miss them. It might be that there is no one to give them back to.

In the end, most will likely stay in European museums far from home. So what should museums do with these cursed treasures? Museums owe it to them to tell their stories. To put these treasures on display and face up to the curse. To allow them to haunt our halls. Can telling their stories lift the curse? Maybe not. But it's all we have.

4 *Haunted Heritage*

The dodo (*Raphus cucullatus*) is the world's most famous animal that has become extinct in the modern age. Anyone in search of the dodo on the Indian Ocean island where it once lived has to make a first stop at the Mauritius Natural History Museum located in the capital city of Port Louis. Immediately upon crossing the threshold of the museum, the visitor sees a sign with "Extinct animals" and a case with bones of the dodo, the Rodrigues solitaire (another extinct bird), and giant tortoises. A brief history of the dodo's extinction is given: it was "encountered in the late 1500s or early 1600s and was probably extinct by the mid 1600s as a result of human activities and especially the introduction of rats and pigs which destroyed the dodo's vulnerable ground nests." The dodo is afforded this central starting position in the museum because of its status as "endemic to Mauritius" and extinct. The presentation aligns with other natural history museums around the world that use

Within the glass case, text on the display plinths reads:

A Unique complete Skeleton of the Dodo
Louis Etienne Thirioux, 1904

Unique Dodo skeleton - Thirioux

Louis Etienne Thirioux born in France in
1846, came to Mauritius in 1870 and
worked as a hairdresser in Port Louis.
Known to be an amateur naturalist he
explored many caves, rivers and hills.
He discovered many bones of Dodo,
Lizards...
His most important find was an almost
complete skeleton from a single Dodo,
... near Pouce mountain1904.

FIG. 28 View of part of the Dodo Gallery of the Natural History Museum, Port Louis, Mauritius, 2022.

the dodo as extinction's poster child. The dodo is often the first animal encountered in extinction-specific galleries or cases.

Farther into the museum, we find the main attraction: the "Dodo Gallery." A whole room filled with dodo stories, images, skeletons, and models. Even knowing that Mauritius was the home of the dodo, the collection is incredibly impressive. It not only includes the scientific presentation of the animal, but also stresses the indigeneity and cultural significance of the bird. Along with scientific drawings, there are also children's drawings. Next to historical artifacts from the settlement of Mauritius, there are modern souvenirs featuring dodos.

The humanities scholarly field known as Extinction Studies has tended to focus on mourning, grief, and guilt as the responses to extinction. The message of these studies is that since extinction is the result of unbridled capitalism, it is something we should be ashamed of. Extinction is understood as an unraveling of multispecies knots of time and generations. But this understanding does not fit for the dodo and Mauritius. Rather than mourning the dodo, the Mauritians seemed to be celebrating it. Rather than unraveling, the extinction seemed to be weaving threads together. This was a different response to extinction than grief or mourning.

How do the extinct still haunt their homelands? We can start to answer this question by examining the iconic dodo on Mauritius and then move to the much lesser-known Japanese river otter. Through these, we explore how extinct species can continue to be felt as real presences through their cultural heritage status. These museum displays narrate the dodo and Japanese river otters as ghosts that still inhabit the landscape of their original homes.

My first encounters with dodos were far from their island home in the Indian Ocean. In these distant settings, dodos represent both extinction and the practice of natural history itself. Dodos are iconic in natural history museums. At the Muséum national d'Histoire naturelle in Paris, the dodo serves as the Extinct and Endangered Species gallery's logo, and at the Harvard Museum of Natural History, the dodo is the logo of the museum even though there is no dodo on display (there was a model in the past, but it is no longer there). Dodos appear in natural history museums more often than any other extinct species (see the appendix for which museums have them on display).

Museums commonly use the dodo to introduce the topic of extinction. For example, the Survival Gallery at the National Museum of Scotland has a dodo model in a freestanding case at one exhibition entrance. The dodo is narrated there as an example of a failure to thrive in the presence of humans:

> As dead as a dodo
>
> The dodo is an icon of extinction. Fat, stupid, and slow, it was inevitable that it became extinct—or was it? Dodos were giant pigeons that evolved on the Indian Ocean island of Mauritius. Without predators, they adapted to island life, losing the power of flight as they became larger.
>
> Dodos appeared stupid because they did not recognize human hunters or introduced animal predators as threats. Dodos were last seen in 1662 and probably became extinct soon after.

In the context of the Survival gallery, which displays how animals have either evolved to thrive or faced extinction, the dodo

FIG. 29
The dodo at the
entrance of the
Survival Gallery
at the National
Museum of Scotland,
Edinburgh, UK, 2020.

is an example of a species that failed the test. The dodo at the
Lee Kong Chian Natural History Museum in Singapore appears in a gallery alongside dinosaurs, comparing the two extinctions by noting that "unlike the dinosaurs, human impact

caused the Dodo's downfall." At the Australian Museum, the dodo appears with this label:

> As dead as a dodo
>
> Why do people say "as dead as a Dodo"? Well, Dodos are extinct and being extinct is about as dead as it gets! You can see a Dodo in this showcase, along with other extinct animals (and some not-so-extinct ones).

Exhibits often use the phrase "dead as a dodo," which is frequently used in similes to indicate that the death is inevitable because of inherent characteristics. There is a common belief (although probably untrue) that the dodo was a stupid slow bird whose nature led to its extinction. The *Oxford English Dictionary* attests the first use of the phrase in 1904. The OED entry for "dead, adj." has a subcategory of phrases: "dead as a door-nail, dead as a herring: completely or certainly dead. Also, (as) dead as the (or a) dodo, (as) dead as mutton." To be "dead as a dodo" must have been well known in the early 1900s, as it appears in a 1918 publication called *Intensifying Similes in English* with the example "Toryism is as dead as a dodo."

Europeans in the seventeenth century knew of dodos: there were some historical accounts of the dodo from the early colonization of Mauritius in the 1600s, and it had been featured in paintings by the famous painter Roelant Savery. A dodo was even seen in London: Sir Hamon L'Estrange recorded that in 1638 "a great fowle somewhat bigger than the largest Turky Cock, and so legged and footed but stouter and thicker and of a more erect shape," was available to see for a fee. Yet the bird was largely forgotten for the next two centuries, until it gained popularity in 1848, when H. E. Strickland and A. G. Melville published *The Dodo and Its Kindred*, the first book-length treatise on the extinct birds of Mauritius and nearby islands. Their book was more than just a natural history of the dodo; they offered a reassessment of all the historical, pictorial, and physical evidence of the bird.

Soon after the publication of *The Dodo and Its Kindred*, a life-size reconstruction of a dodo appeared at the Great Exhibition in London in 1851, and the Oxford University Museum of Natural History put its dodo remains (the only fleshy remnants of the species) on display in 1860. The latter truly cemented the dodo in the public mind. It was these dodo artifacts that inspired Charles Dodgson (writing under his pen name Lewis Carroll) to craft a dodo character in *Alice in Wonderland*. The masterful accompanying illustration by Sir John Tenniel became the standard image of the awkward bird. Franco-English writer Hilaire Belloc published a poem in his wildly popular *The Bad Child's Book of Beasts* (1896) about his own encounter with the Oxford dodo on display:

> The Dodo used to walk around,
> And take the sun and air.
> The sun yet warms his native ground—
> The Dodo is not there!
> The voice which used to squawk and squeak
> Is now forever dumb—
> Yet may you see his bones and beak
> All in the Mu-se-um

Belloc's observation was that dodos no longer walk around on Mauritius to be encountered, but are instead found in the museum.

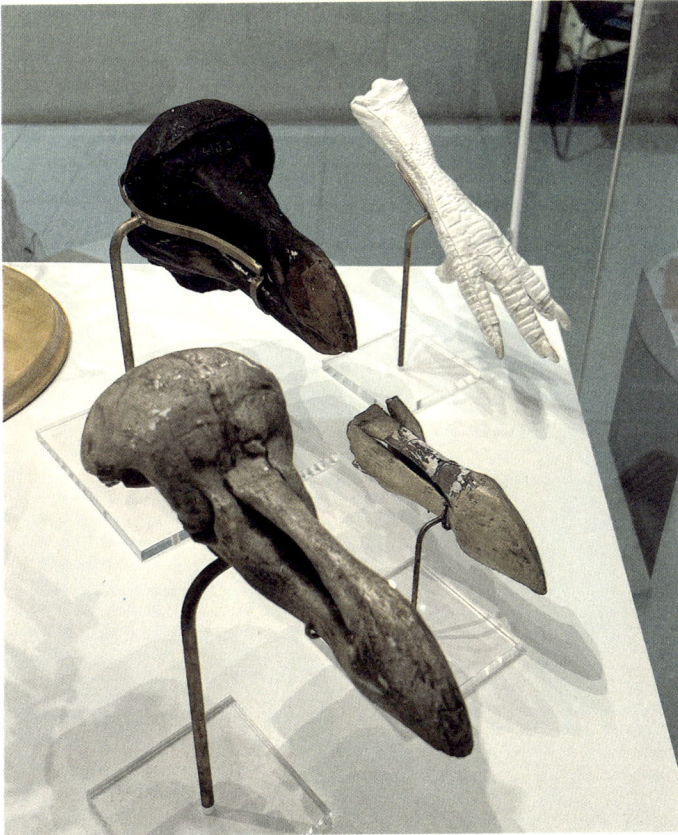

FIG. 30
Plaster casts of the few
bones from modern
dodos, including the
Oxford head, at the
University Museum of
Zoology, Cambridge,
UK. The case is one
of the first things the
visitor encounters in
the museum, 2022.

In museums across the globe, dodos are displayed as bones, composite skeletons, wooden models, faux taxidermy, and other models. The skeletons are all subfossil composites excavated from the Mare aux Songes swamp on Mauritius. Archeological digs in the swamp have provided a veritable trove of bones, but these are typically a jumbled mess with few intact skeletons. Therefore, skeletons for display are created with bones from various individuals stuck together. These dodo bones are not recent—they are about 4,000 years old, rather than from when humans contacted the bird 400 years ago. Naturalist Richard Owen's famous dodo in the Natural History Museum in London, which we encountered in the treasures room, is this kind of dodo. There is a dodo skull in the Statens Naturhistoriske Museum in Copenhagen that is not a subfossil and, as we have seen, it is kept in a special alarmed vault to protect it as a rare specimen. Importantly, there are no stuffed dodos; the only fleshy remains are a single mummified head kept by the Oxford Natural History Museum. There was previously a foot in London, but it has been lost. Molded copies of the Oxford head and London foot (the mold was made before it was lost) have been purchased by many museums.

Faux taxidermy of life-size dodos are also commonly displayed. The feathers of dodos don't exist, except for one small feather recovered from the Oxford dodo head; therefore, feathers from other birds like turkeys and geese are used to construct dodos for display. These feathers give the models a life-like appearance. Watching the crowds taking photos and talking about the dodo at the Natural History Museum in London, it is clear that visitors believe they are seeing a stuffed bird. The dodo they are seeing is a *real* artifact—but it doesn't contain any genetic material of a dodo. It is a representation of a dodo, and a contentious one at that.

While all dodos we see in museums are scientific constructions, only some displays emphasize the changing idea of the dodo. Over time, scientists have discovered more and more about the species, which has existed as a shadow with unclear outlines. The evidence remaining from direct contact with dodos when they were still alive, such as Savery's art and the Oxford head, is thin, so scientists have made conjectures. These required updating as more skeletal remains were found

on Mauritius and as artistic evidence was reevaluated. A special exhibit on evolving dodo models called "Dodos: Old Bird, New Tricks" at the Natural History Museum at Tring focused on the changing scientific view of the dodo based on new skeletal finds in Mare aux Songes and the reexamination of existing bones. In exhibits like this, it is common to use pictorial evidence along with natural history specimens to present what we know about an extinct species. New ideas about the dodo's body configuration are also on display at the Senckenberg Natural History Museum in Frankfurt. The exhibit text stresses that this dodo combines "representations from the 17th century and the latest scientific findings." In particular, the dodo's tail size and overall coloration are a "realistic representation" based on a seventeenth-century Indian painting and the powerful legs and decurved bill were based on new skeletal studies.

FIG. 31
The earliest illustration including a dodo (on left next to a turtle) in a copy of *Het Tweede Boek* published in 1601. This copy belonged to museum founder Walter Rothschild. On display in the "Dodos: Old Bird, New Tricks" exhibit, Natural History Museum at Tring, UK, 2017.

In dodo presentations outside Mauritius, the dodo is an object of scientific study, the epitome of extinction. But the dodo in these contexts is never a cultural creature. Artistic works are mobilized as evidence of the body size or color, but not evidence of the dodo's history and interaction with humans. This is where the dodo on Mauritius would prove very different.

EXTINCTION AS HERITAGE

Passing under the doorway with "Dodo Gallery" written above, the visitor in the Mauritius Natural History Museum is greeted by three skeletons in glass. The most unique of these is the closest to the entrance—the world's only complete dodo skeleton. Whereas many museums have composite skeletons constructed from the remains of many birds put together to fill in the gaps, this was one individual dodo that lived and died near the Pouce Mountain of Mauritius. The specimen was found in 1904 by amateur naturist Louis Etienne Thirioux and takes pride of place in the museum. This unique specimen is supplemented by others in the room.

Also in this center display are two other extinct endemic birds: a taxidermy mount of the Mauritius blue pigeon (*Alectroenas nitidissima*) and a skeleton of the red rail (*Aphanapteryx bonasia*). The blue pigeon was named after its dark-blue wing feathers, but the feathers on this specimen have lost their blueish tint and now appear a dull dark brown. The only other blue pigeon I've seen is at the National Museum of Scotland, and it has better-preserved plumage. The individual in the "Dodo Gallery" was the last known blue pigeon, killed in 1826. Standing in the case next to it is the red rail, a flightless bird with reddish-brown feathers and a long beak, which exists today only as

FIG. 32 (*facing*) Unique complete skeleton of a dodo on display in the Dodo Gallery, Natural History Museum, Port Louis, Mauritius, 2022.

FIG. 33
Mauritius blue pigeon
in the Dodo Gallery
of Natural History
Museum, Port Louis,
Mauritius, 2022.

bones. There are no taxidermied specimens of the red rail anywhere; it is possible that an illustration of the red rail made in
1610 is based on a taxidermy mount, but if so, the specimen
has been lost or destroyed. The red rail was regularly caught
and eaten by sailors stopping on Mauritius in the early 1600s.

After the dodo became extinct around 1662, some sailors may have confused the red rail for the dodo, so late 1600s accounts mentioning dodos are likely describing red rails.

But more than a natural history, this room presents a cultural one. The opening interpretive panel makes that clear: "The dodo is the national emblem of Mauritius: bank notes, coins, stamps, matchboxes and a host of objects in ordinary life bear the image of the dodo. It is part of the Mauritian cultural heritage as much as the panda is a symbol of China." Going clockwise around the room, three information boards begin with standard natural history information about the dodo, like height, weight, and diet, but the dodo's status is highlighted in the first board's title, "Extinct for over 350 years!" Early sketches of the dodo accompany text about the earliest documents attesting to the species. Finally, the boards present the first archeological investigations from 1865, which resulted in a huge quantity of dodo bones being shipped to Europe.

Then the visitor encounters the history of colonization of Mauritius. From 1598, when Dutch ships began stopping at the island in search of food, "the dodo was doomed. . . . Extinction was inevitable." Extracts of the eyewitness account published in 1601 are used to annotate the reproduction of a drawing of the island's forest and its use by sailors. The cases here have cultural artifacts like chinaware and traded goods.

The display then turns to the cultural history of dodos in museums. The far wall has three display boards on the history of finding the unique complete skeleton; the dodo's diet as described in historical texts; and the history of dodos taken abroad by sailors to Europe and Asia. The next three are about the history of museum remains, including lost specimens and the only soft tissue remains, which is the Oxford head. At the far end, there is a dodo model and an Artificial Reality (AR) reconstruction that adds a moving dodo parent and squawking dodo chick to the scene (more will be said about the AR in a later chapter). Labels provide information about how models of the dodo's body configuration were recently updated based on the Mare aux Songes subfossil bone discoveries.

Finally, there is a case of cultural ephemera featuring dodos, from ceramic plates to figurines, keychains to artworks. Drawings of dodos from scientific reconstructions and children's drawings are hung in the room. This takes the dodo from a historical animal to an animal in the present. What is obvious in this room is that the extinct dodo is still very lively. It has a cultural resonance that makes it more than a museum specimen.

The cultural importance of the extinct dodo becomes more apparent in the cultural history museums of the island. Take for example the Frederik Hendrik museum, which recounts the history of the Dutch fort on the site. The exhibit in the interpretive center discusses the arrival of the Dutch on the island in 1598 up to their departure in 1710. When approaching the exhibition, however, the visitor is not faced with Dutch merchants, but rather a dodo statue as the centerpiece. The displays include information on the dodo in Dutch texts and drawings. In this display, the text argues that the Dutch were not directly to blame for the dodo's extinction:

> There is a common belief that the Dutch killed the dodo. However, this bird was known to them as walghvogel (disgusting bird), because it tasted very bad. It is now believed that pigs and rats were much more damaging to the dodo population. This is consistent with the fact that not a single dodo bone has been found during the excavations.

The retelling of the cultural history of the Dutch period is framed by concern for the dodo. In the presentation of the second period of occupation, 1664–1710, "the last dodo" mentioned in 1688 in the diaries of the commander of the Dutch forces gets a large display. Drawings of birds from the islands that became extinct as a result of colonization are hung around the room with interpretive labels highlighting the human causes of extinction: the dodo, Rodrigues solitaire, Mauritius Dutch (blue) pigeon, and red rail. At the Frederik Hendrik museum, in a history that might have been presented without reference

FIG. 34
The dodo in the center of the Dutch history on Mauritius exhibit at the Frederik Hendrik museum, Mauritius, 2022.

to the environment, the extinction story of the dodo is the major narrative thread.

The National History Museum in Mahebourg likewise includes the dodo story within the history of Dutch colonization. A case of dodo bones sits under a reproduction of the famous painting of the dodo by Savery (1650). The label notes that the dodo, along with other birds like the red rail and blue pigeon, "become extinct in the 17th century during the Dutch settlement in Mauritius 1638–1710" due to "1. introduction of foreign animals—pigs, rats, goats, deer, monkey and cattle; 2. destruction of their natural habitats; and 3. hunting." There is no hedging about who is to blame for the extinction of the dodo. The history of the settlement of Mauritius, which was an unpeopled island prior to Dutch colonization, is a history of extinction, so celebrating one means acknowledging the other.

The dodo at home is different than the dodo in museums in the rest of the world. Whereas the dodo is a symbol of extinction writ large in museums across the globe, on Mauritius, it is a living part of a cultural history. It inhabits the halls of museums to haunt the colonization narrative and cast a shadow over it. The settlement of Mauritius starts with loss, and this is not brushed over.

EHIME'S ICON

While the dodo is known the world over as an icon of extinction, the Japanese river otter (*Lutra nippon*) is known to very few. *L. nippon*, which is a separate species of river otter from the common Eurasian otter *Lutra lutra*, was declared extinct in 2012 by the Japanese Ministry of the Environment. It ranged in rivers across Japan centuries ago, but in the last hundred years was found primarily on Shikoku Island in southern Japan. It was a victim of pelt hunting and habitat destruction, leading to a rapid decline and then extinction in the 1960s. Yet it is still the official animal of Ehime Prefecture, which is adjacent to Kōchi Prefecture on Shikoku, where the last river otter was seen in 1979.

The Ehime Prefectural Science Museum in Niihama, Japan, is the place to encounter the Japanese river otter. It has the largest collection of Japanese river otter remains anywhere. The first display case with the animal is a typical display on extinction and endangered animals that includes three river otters mounted in various poses. One *kawauso*, as the Japanese river otter is called in Japanese, is rather playful, standing on its hind legs. The large label with these three otters noted that the *kawauso* was listed as a Special Natural Monument of Japan in 1965. This designation is given to animals that "need to be preserved as well-known characteristic Japanese animals" in the Law for the Protection of Cultural Properties. The Agency for Cultural Affairs can designate animals, plants, geological features, and ecosystems as natural monuments under the law. A locally extinct butterfly (*Fixsenia iyonis*, also known as *Strymonidia iyonis*) alongside the otters was designated a Prefectural Natural Monument in 1962. A Japanese dormouse on display is another Special Natural Monument, but it is not yet extinct. Denoting a species as a natural monument by a cultural affairs department marks these species as culturally important, not just scientifically important.

While a case with three Japanese river otters along with other endangered and extinct species is impressive, turning the corner to the Ehime ecosystems gallery, the visitor finds the museum's real river otter display. The section "The Field and

Mountain of Ehime" features a Japanese river otter as its symbol and displays the animal in various ways: a diorama with four model animals, three taxidermy mounts, one skeleton, one flat pelt, and significant information boards. The scale of the exhibit as a single species display is matched only by the

Dodo Gallery in the Mauritius Natural History Museum and the thylacine room in the Tasmanian Museum and Gallery in Hobart. There is something different about displaying an extinct species in its home territory contra a faraway museum. In its native location, the number of specimens can abound and the animal takes on a more personal meaning.

The diorama is a particularly significant element of the display. Habitat dioramas place taxidermy or model organisms into a scene representing their original habitat. These are a form of "ecological theatre," according to historian Karen Wonders. Dioramas developed in the United States and Sweden in the 1880s to 1930s as a way of bringing wilderness landscapes to the eyes of an urban public distanced from it. While dioramas were based on scientific knowledge, their inviting form imitating nature moved displays away from stale collections of dead scientific specimens for elites to seemingly alive educational exhibitions for public instruction. The aesthetics of the habitat diorama are integral to the experience of the display because "they extend beyond the individual specimens themselves to the sensual impression created by scenery as a whole," according to Wonders.

While habitat dioramas became commonplace in natural history museums, especially after the Second World War, extinct species are almost never put into them. In all of my travels, I have found only a few habitat dioramas in which the specimens are in a contained landscaped scene with a background painting to simulate observing a complete landscape. Most of these are birds: Labrador ducks on a snowy shoreline at the American Museum of Natural History in New York; a great auk on a rocky coast at the Naturhistorisk museum in Oslo; and passenger pigeons in the forest at the Academy of Natural

Sciences and in an autumnal Iowa woodland in the Denver Museum of Nature and Science. Even a smaller number of dioramas have extinct species alongside others: there is a reconstruction of Avery Island, Louisiana, with both ivory-billed woodpeckers and Carolina parakeets in Denver; a southern swamp that includes a model of a Carolina parakeet along with taxidermy mounts of still-living species at the Illinois State Museum; and a collection of northern birds (great auks, Eskimo curlew, and Labrador ducks) on an icy ledge at the Academy of Natural Sciences. The only mammals other than the otters in a diorama are Japanese sea lions: a group of three taxidermied animals in the Shimane prefectural natural history museum (the visual appearance of this one is greatly hindered by a recently installed foldable plexiglass screen) and a model of a bull sea lion in the Ulleungdo Marine Protected Area Visitor Center, South Korea.

The diorama in Ehime features four Japanese river otter models on a rocky shoreline with one about to dive into a tidal pool. The scene seamlessly blends the three-dimensional rocks into the background painting to evoke a full landscape. The otters appear playful and vibrant. Yet, these are models, not biological specimens, so they never were alive. The diorama is open to the room's atmosphere—the glass only extends from the bottom about a quarter of the height, likely to keep children's fingers away—so it was probably not seen as secure for such precious biological specimens. The museum does, however, own a plethora of them. There is a list of all their specimens as part of the display to the right of the diorama, which shows a whopping thirty-five otters collected between 1954 and 1971. The list notes how each otter died: they were often found dead or accidentally killed, but sometimes the cause is

FIG. 36
The Japanese river otter
diorama with models
in an ecosystem scene,
Ehime Prefectural
Science Museum,
Niihama, Japan, 2023.

unknown. The location of the collection of the specimens is shown on an adjacent map. This is the only exhibition I have encountered that lists in this way all the specimens that the museum holds of an extinct species.

In front of the diorama, the permanent sign features a drawing of an otter with biological information and text about its history: "They used to live all over Japan but since the Meiji period, they have been overhunted and their habitats lost to development. Extinction is feared." But pasted below this is a label adding that the Ministry of Environment announced on August 28, 2012, that the otter was designated as an extinct species because a sighting had not been confirmed for thirty years. The diorama had been in place before the otter moved from endangered to extinct, so the information had to be updated.

The floor space given to the display of the Japanese river otter demonstrates its significance to Ehime. No wonder the mascot of Ehime Prefecture is Gen-chan the Japanese river otter (the towns of Ainan in Ehime and Susaki in Kōchi also have Japanese river otter mascots). There is a local specificity about the display, with the list and map showing visually the links between the animal and the local geography. While there is a skeleton of the otter on display in the National Museum of Nature and Science in Tokyo, this is put into an "Endangered Organisms" display with a small label along with a Honshū wolf skeleton as a warning about extinction for other still-extant Japanese species. Such a choice is very different from what the Ehime museum has done to claim the Japanese river otter as its own.

CLAIMING GHOSTS

The dodo and the Japanese river otter are both claimed as valuable former inhabitants by local museums. Both are portrayed as part of the national heritage—the otter with its official heritage status and the dodo with its proliferation in Mauritian history and culture. The dodo on Mauritius and the Japanese river otter in Ehime are not the norm. While these two locations have specimens of their extinct animals, many other animals' bodies exist only in colonizer collections housed in European museums. This makes the Mauritius and Ehime displays different than the standard extinction displays that present animals from faraway places. The dodo and Japanese river otter in their places of origin are embraced as part of the local cultural heritage.

Exhibitions in these local settings are able to tell much more in-depth and situated stories than their counterparts in museums that do not have the personal connection. It's not just that they have more specimens in their collections; it is that they place those specimens into the local landscape. The dodo is physically present on Mauritius through displays of the early Dutch drawings of the dodos in their habitat and the Mare aux Songes archeological investigations. The Japanese river otter is placed in the Ehime landscape through a habitat diorama, maps of specimen collection, and local mascots. Because extinction is a locally embedded issue, encountering the remains of an extinct animal in the place it used to live has particular poignancy. While the stories of their extinctions are tragic, there is no shame in these presentations. Instead, these extinct animals are celebrated.

5 *Last Remnants*

I met Martha, the last passenger pigeon, in 2015. Martha had died in 1914, but I saw her as part of a commemoration of the 100th anniversary of her death. Perched on a branch with her body facing away from me, her head was turning in my direction, looking at me with a glassy red eye. In this exhibition, she was placed near a male bird reaching out with a seed in his beak, but it was not her previous mate, George. In front of Martha and the male, a passenger pigeon skin specimen lay belly-up, as a sure sign of the death of the passenger pigeon. This bird was the opposite of Martha, who somehow still looked lively.

Martha was the last of the passenger pigeons, at least as far as we know. This means that she was the *endling*, a neologism coined to represent the last individual of a species. *Endling* doesn't appear as a word in standard dictionaries, but it has gained currency in the twenty-first century. It started in April 1996, when the well-respected journal *Nature* published a letter

FIG. 37 Passenger display of "Once There Were Billions" special exhibition at the Smithsonian. Martha, the last passenger pigeon, is on the left. Smithsonian Museum of Natural History, Washington, DC, USA, 2015.

from two men working at a convalescent center proposing that a new word be adopted to designate a person or individual of a species that is the last in the lineage: *endling*. They had patients who were dying and thought of themselves as the last of their lineage, and they needed a word to use when discussing it with their patients. The suggestion of *endling* was met with counter-suggestions in the May 23 issue of *Nature*: *ender* (Chaucer used it to mean "he that puts an end to anything"), *terminarch* (because it has a more positive ring than *endling*, which sounded pathetic, according to the respondent), and *relict* (which means "last remaining," but typically for a group). Nothing more appears to have been made immediately of the suggestion or the word in scientific circles.

But the word got a boost from Down Under in 2001, when the National Museum of Australia (NMA) opened its doors in Canberra. NMA featured a gallery called Tangled Destinies, and the world *endling* appeared. On the wall behind two thylacine specimens was written a definition: *Endling (n.) The last surviving individual of a species of animal or plant.* The curator Mike Smith had seen the *Nature* letter and thought that *endling* was the right name for the last of an animal species. Although these two specimens were actually not the last thylacines, the word evoked a melancholy tone of loss for the exhibition. Since that appearance, the word has slowly seeped into popular culture, appearing in symphonic music, performance art, science fiction stories, comics, and other artworks. Through 2020, the shiny metal display box with the definition was still in the gallery—the endling continuing to draw visitors into relationship with the thylacine skeleton and mummified head in the room. It has, however, since been removed and the gallery completely redesigned.

How does the endling, the last individual of a species, haunt extinction stories in museums? In this chapter, we venture into exhibitions to uncover how the endling haunts narratives of the passenger pigeon and great auk. In light of these hauntings, we can also consider how de-extinction attempts are destabilizing the narrative possibilities for the quagga's end.

THE PASSING OF PASSENGER PIGEONS

When John James Audubon wrote about passenger pigeons as an accompaniment to his drawing of life-size birds on an oak tree branch, he remarked that "the multitudes of Wild Pigeons in our woods are astonishing." In 1813, he had encountered a gigantic flock in Ohio: "The air was literally filled with Pigeons; the light of noon-day was obscured as by an eclipse, the dung fell in spots, not unlike melting flakes of snow; and the continued buzz of wings had a tendency to lull my senses to repose." While this sounds like an idyllic nature encounter, Audubon goes on to describe the buzz that the pigeons created in the community: "The people were all in arms. The banks of the Ohio were crowded with men and boys, incessantly shooting at the pilgrims, which there flew lower as they passed the river. Multitudes were thus destroyed. For a week or more, the population fed on no other flesh than that of Pigeons, and talked of nothing but Pigeons."

It was the passenger pigeon's usefulness that led to its slaughter. When they passed overhead, the plate would soon be passed full of food. Pigeon pie was a common dish for the masses. Audubon had not thought that the killing of passenger pigeons, even in great numbers, was affecting the population—"no apparent diminution ensues" was the way he put it. But only eighty years after those words appeared in print, there was only one known passenger pigeon left.

The immense number of passenger pigeons is impossible to capture in a museum exhibition. The American Museum of Natural History in New York has a large collection on display— ten or so birds make up a flock perching in an oak branch and standing among the leaves littering the ground as "Passenger pigeons in autumn." The label has to do the work of stressing the abundance, noting that naturalist Alexander Wilson estimated a flock he saw in Kentucky in 1808 to be 2,230,000,000 birds.

The decline of passenger pigeons had been rapid and seemed to catch commentators off-guard. Between 1894 and 1898, the

FIG. 38
Passenger pigeon group at the American Museum of Natural History, New York, USA, 2022.

local bird expert Oscar Byrd Warren of Palmer, Michigan, solicited reports about passenger pigeon sightings. The people who wrote back always noted that the previously common birds had not been seen in a few years, but did not know why. In their letters, informants always assumed that the pigeons had moved to somewhere else. For example, H. T. Blodgett of Ludington, Michigan, wrote in 1904 that "much to my regret I have seen none of the beautiful birds for about six years. The savage warfare upon them, from nesting place to nesting place by pothunters and villainous fellows who barreled them for market, with nets and every brutal means for wholesale destruction, has driven them, I know not whither." Observations of passenger pigeons did trickle in, including a sighting of a flock of about fifty in southern Missouri in 1896, and approximately 300 birds in Wisconsin in 1897. Although one writer admitted that in the Midwest states "large flocks of Passenger Pigeons are a thing of the past," small flocks and individual sightings were still recorded. While everyone acknowledged that the bird's numbers had greatly decreased, they did not assume that it was on the verge of extinction.

Yet by the 1910 annual meeting of the American Ornithologists Union, the dire situation for the passenger pigeon was recognized. A reward of $300 was offered by Colonel A. R. Kuser for information leading to a pair of undisturbed nesting passenger pigeons, and others issued similar rewards. The idea was to start a "Passenger Pigeon Restoration Club" if some live birds could be located. C. F. Hodge of Clark University claimed that "negative evidence is proverbially inconclusive" and still held out hope that a couple of seasons of searching would be needed to decide whether or not to abandon the search. The search was continued into 1912 because a number of "apparently encouraging reports had been received," but no confirmation of passenger pigeons was ever found. There were no more passenger pigeons left in the wild.

However, there were some in captivity. Martha the pigeon had lived her whole life in captivity, first as part of a breeding project by professor of zoology Charles Otis Whitman, then in an aviary at the Cincinnati zoological gardens in Ohio. She had lived alone since 1910, when her mate had died. That made her the last, and the zookeepers knew it. So when she died September 1, 1914, her body was packed on ice and shipped to the National Museum of Natural History (now the Smithsonian) in Washington, DC. Martha was prepared as a taxidermy mount, and in 1932 the Smithsonian put her on display in the National Museum's Bird Hall with a biographical label: "MARTHA last of her species, died at 1:00 p.m. 1 September 1954, age 29, in the Cincinnati Zoological Gardens. EXTINCT." Later she was moved to the Birds of the World exhibit, where she stayed until it was shut down in 1999 to make room for a new mammals hall. She was relegated to storage and did not come out until 2014 to appear in the *Once There Were Billions* exhibition to mark the centennial anniversary of her death and the extinction of the passenger pigeon. After that exhibit closed, she moved into the Smithsonian's Objects of Wonder exhibition of treasures, taking her place among a treasures gallery. The story of the passenger pigeon's extinction became centered on Martha's story.

MARTHA'S HAUNTINGS

Passenger pigeons are remarkably common in natural history museums. Passenger pigeon bodies—either taxidermy or study skins—are on display at over twenty museums that I visited. When they were alive, they were the most numerous birds on the planet, so it's no wonder that museums had been able to acquire them.

The story of the end of passenger pigeons with Martha is repeated over and over again in natural history museums around the world. The story is about the *last* (or its equivalent in other languages, such as *dernier* in French), a word used in almost every single passenger pigeon label. Sometimes the last individual goes unnamed—for example, the label at the Canadian Museum of Nature just says: "the last Passenger Pigeon died in captivity in 1914 at the Cincinnati Zoo in Ohio, U.S.A." But Martha often gives a personal dimension to the story: a name and a date. The Natural History Museum at Tring has a typical label in this regard: "it is estimated that there were three to five billion passenger pigeons when Europeans discovered America. The last captive bird, Martha, died in Cincinnati Zoo in 1914."

Martha's story haunts, and even overshadows, the extinction. Take, for example, the National Museum of Scotland, which has a pair of pigeons mounted on a branch in their extinction display case in the Survival gallery. Pasted onto the glass in front of these birds is a semi-opaque label with the title "Martha, the last passenger pigeon." It features a large photograph of her with a paragraph on the numerousness of the species, followed by a statement about Martha's end: "The last bird, called Martha, died at exactly one o'clock in the afternoon on the first day of September 1914 in Cincinnati Zoo." The precision of the statement (whether or not it is accurate) stresses the finality of Martha's death as a historical event. Martha's personal story comes to stand in for all the pigeons on display.

The exhibition "Martha's legacy" housed in the Cincinnati zoological gardens aviary pavilion where she died over a century ago manifests her ghostly haunting. The "Passenger pigeon memorial" is marked on the zoo map. When you get there, you see that there is a small memorial marker outside a building with a brass pigeon statue, but this is not the memorial—the building itself is the memorial. Beyond the big heavy oak open doors hangs a large print of an amazing, much larger wall mural. A flock of passenger pigeons, the males with bright blue heads and red chests and the females with vanilla and taupe feathers, streams across the scene from right to left. Those flying from the trees in the distance are being joined by a group escaping through the closed gates of the zoo's pavilion. One female, Martha, leads the ghostly flock on their last flight. The mural "Martha, the Last Passenger Pigeon" was designed by John A. Ruthven and executed on a gigantic building wall abutting a parking lot in downtown Cincinnati with the help of youth apprentices in 2013 as part of the extinction centennial. "Martha's legacy" is framed here as lessons for sustainability. On the left, the exhibit narrates the passenger pigeon's decline from billions to none, with a special emphasis on Martha as the very last. On the right, the exhibition interprets the disappearance of the passenger pigeon as "A wake-up call to save wildlife," recounting some historical and contemporary conservation projects, especially those the zoo is involved in. This is the hoped-for legacy of Martha.

Although the Cincinnati zoo claimed a conservation victory with her death, many species discussed in this book have gone extinct since the passenger pigeon. How many birds will be on display in the exhibit for the 200th anniversary of Martha's end in 2114? Lessons obviously have not been taken to heart.

FIG. 39
Passenger pigeon on
display at Wollaton
Hall, UK. Below it is a
label that starts "Once
the most abundant
North American bird
with a population of
3 to 5 billion. The last
one, called Martha,
died in Cincinnati Zoo
in 1914," 2022.

INSIDE OUT

I've seen many bodies of extinct animals in museums across
the world, but I have never felt a more profound sense of sor-
row than during a visit to the Statens Naturhistoriske Museum

in Copenhagen, Denmark, in December 2021. Four jars stood on a shelf, filled with a sickly yellow fluid containing ghostly white forms. On the shortest of the jars, a label is visible: "Alca imprennis. Island 1844. (♂)." A stuffed great auk with an egg stands in front of the jars—but it was collected from 1832, when there were still other living auks. The ones in the jars had only each other. According to the signage, these are "the guts and eyes of the world's last breeding pair, caught on the Icelandic island Eldey in 1844."

There was something deeply unsettling about these particular specimens. There was no illusion like in taxidermy of them being whole. Instead, they were pieces of the inside of the birds taken out of their bodies. These jars encapsulated death—a violent, inexcusable death of a species. They had been turned inside out.

The great auk's story is a great tragedy. The great auk had once roosted on rocky islands of the North Atlantic from Newfoundland to Norway. A large, flightless seabird, it was a powerful swimmer but slow and awkward on land when it came ashore a few weeks a year to nest. It was during the nesting phase that it became vulnerable to human predation. The birds were big, with lots of meat, and the eggs were easy targets. The populations of great auks on the western side of the Atlantic were hunted for meat by early European settlers and the populations disappeared, leaving the only significant roosting population in Iceland's icy waters by the end of the eighteenth century.

From the mid-eighteenth century, European upper-class naturalists discovered the great auk as a prize they needed for their egg and skin collections. This turned what had long been a subsistence hunt into a commercial one. Sailors could make the treacherous voyage out to the rocky island rookery and come back with a bag of great auks. They would skin the birds and eat the meat. They would make a hole in the egg and blow out the contents to eat, while keeping the eggshell intact. In both cases, they ended up with a saleable product (a skin or an egg) that was in high demand from merchants in Reykjavik for the collectors' market on the continent.

The eggs of the great auk are indeed beautiful. They are oblong with one pointy end. Each has a unique pattern of black markings on a cream background like a Jackson Pollock painting. These eggs are rarely on display, but have appeared in the special exhibitions *Once There Were Billions: Vanished Birds of North America* at the Smithsonian and *Nature's Treasure Trove—200 Years of Naturalis* at Naturalis, Leiden. One is on permanent display with a guarding adult below the jars with the last great auks in Copenhagen. But it isn't the last egg—that one was smashed and not collected.

In Gísli Pálsson's book on the great auk and its extinction, he recounts the hunting trips that zoologist Alfred Newton and egg collector John Wolley heard about when they visited Iceland in 1858. Newton and Wolley's *Gare-Fowl Books*, which served as their diary and interview notes for the trip, record a sad tale of continued annual exploitation even when numbers were vastly dwindling. In 1830, the island on which most of the auks were nesting, Great Auk Skerry, sank due to a volcanic eruption. There was, however, still a very small population nesting on the island of Eldey. It was the last population of great auks, although the local Icelanders didn't realize that at the time. The hunters changed their target to Eldey: they found dozens of adults in 1830, bagged sixteen in 1832, and took another twenty-four in 1833. One merchant sold eight birds and

eight eggs in 1834. The last successful trip to hunt great auks was in 1844, when two birds were killed. These last two were sold by the leader of the hunting party, Vilhjálmur Hákonarson, to speculator Christian Hansen, who then sold them onward. In this particular instance, the organs were not thrown away prior to sale, and these are the organs that ended up in the alcohol jars in Copenhagen.

There are more jars containing other organs of these two last great auks in the back room of the museum, but the four on view are enough. There is something eerie, haunting, about looking at the insides of something dead. The taxidermied great auk standing in front of the jars is just as dead as is the egg at its feet. But it doesn't *feel* quite as dead. Wet preparations of extinct species are rarely on display—I saw one thylacine pup at the Australia Museum, one thylacine skin at the Grant Museum of Zoology, and the Jamaican giant galliwasp in Glasgow. Copenhagen's museum is an exception, as it seems to revel in its wet specimens: a case packed with fluid-filled jars containing Charles Darwin's crustacean collection, the only specimen of the spider *Pardosa danica* that exists, and five glass jars with the giant anoles *Anolis roosevelti*. The anoles are probably extinct. Gitte Westergaard has traced the discovery of the species on the island of Culebra of Puerto Rico, the movement of the species to the museum, and its legacy as an object of conservation on Culebra. The anoles in these jars, which had entered the museum collection in 1863 but were lost and not rediscovered until 1986, broadened the known range of the lizards from Culebra to nearby Caribbean islands. They stand now awkwardly in their fluid jars, all color faded away like an apparition. Seeing specimens in preserving fluid is an encounter with fleshiness that isn't quite right. It is unsettling to be face-to-face with the gooeyness of ghosts.

It is both unsettling and emotional, and particularly sorrowful. The loss of species and disappearance of ecosystems, or even the anticipation of their loss, can manifest itself emotionally as ecological grief, as scholars such as Joshua Trey Barnett, Ashlee Cunsolo, and Neville Ellis have shown. But, as Barnett observes, it is not inevitable that an extinction will lead to grief, especially if the person has difficulty recognizing nonhuman losses as losses. The communication rhetoric—the objects put on display, their positioning, the interpretive text, the room, and even the gallery itself—deployed at Statens Naturhistoriske

FIG. 42
Two of the five giant
anoles on display,
Statens Naturhistoriske
Museum, Copenhagen,
Denmark, 2021.

Museum enables grief. The museum starts by asking the visitor to value nature through the naming of the exhibit: *Precious Things*. The visitor wanders through the gallery, randomly encountering wonders of nature. The lights are dim, creating an air of both solemnity and mystery. The choice to put the organs of the last great auks on display, rather than simply showing visitors a stuffed bird and egg, aids in making the loss visceral. The interpretative sign with its yellowish lettering on a dark-gray background does not use the word *extinct*, but it does start the story with a "there was once" formulation—"The great auk was once . . ." (Gejrfuglen var engang)—a signal to the reader of different times in the past. The text makes it clear that the great auk was hunted for food then hunted by collectors and museums, and that what the visitor sees in the jars belongs to the last breeding pair.

Perhaps the emotion of grief seeing the auks in those jars is not that different from Aldo Leopold's emotion when the extinct passenger pigeon was being honored with a monument in 1947:

> We meet here to commemorate the death of a species. This monument symbolizes our sorrow. We grieve because no living man will see again the onrushing phalanx of victorious birds, sweeping a path for spring across the March skies, chasing the defeated winter from all the woods and prairies of Wisconsin. . . . There will always be pigeons in books and in museums, but these are effigies and images, dead to all hardships and to all delights. Book-pigeons cannot dive out of a cloud to make the deer run for cover, nor clap their wings in thunderous applause of mast-laden woods. They know no urge of seasons; they feel no kiss of sun, no lash of wind and weather; they live forever by not living at all.

BACK FROM THE DEAD

The last quagga, like the endling thylacine, passenger pigeon, Carolina parakeet, and Pinta Island tortoise, died in a zoo. On August 12, 1883, the last known quagga died in the Artis Zoo

in Amsterdam. I encountered its mounted body on display in the *Nature's Treasure Trove—200 Years of Naturalis* special exhibit in Leiden. Like all the specimens on exhibit there, it shone under the lights in front of a black background. The alternating white and reddish-brown stripes starting at the eyes and continuing down the neck popped out. The colors of its mane matched the alternating hide.

FIG. 43
The last quagga on display in the *Nature's Treasure Trove—200 Years of Naturalis* special exhibit at Naturalis, Leiden, the Netherlands, 2021.

But even though Naturalis presented this as "The last quagga" in its label title, the text did something else. The first paragraph about the death of the animal in the Artis Zoo ends with "Its death marked the extinction of the quagga . . ." Those dot-dot-dots are key because of what happens in the next paragraph: "We now know that the quagga was not a separate species, but a subspecies of the common zebra. Through selective breeding, scientists are in fact today able to re-create a quagga with its red-brown coat." The label turns the story of the endling on its head in two ways: first, that it wasn't really a species, and second, that it has been re-created.

Species are notoriously difficult to define. At a base level, a species is a group of organisms that are all broadly similar and can breed with each other and produce fertile offspring. Before the advent of genetic testing, morphology (i.e., the animal's body configuration) or phenotype was the deciding factor about whether something was a new species. Morphology is still the most widely used basis for describing a new species. If it has features that other species do not, or is missing things they do have, then it is understood as something new. The quagga, which modern scientific nomenclature names *Equus quagga quagga*, was named by the Dutch naturalist Pieter Boddaert as *Equus quagga* in 1778, placing it as a new equine species (equines include horses and zebras). Quaggas and zebras were differentiated by the basic differences in striping patterns. In the 1980s, the DNA of quagga analyzed from museum specimens was found to be closely related to plains zebras. Genetic studies in the 2000s concluded that the quagga was a subspecies of plains zebras, which tend to have less striping the farther south they live. The modern designation of the quagga as a subspecies privileges genes as the deciding factor for a species, rather than visual characteristics.

Interestingly, however, the de-extinction of the quagga—the second claim on the label—focuses precisely on the visual characteristics of its coat. While it may sound like science fiction, de-extinction, which is the remaking of an extinct species by either backbreeding or advanced genetic manipulation, is an ongoing scientific pursuit. One of the earliest targets of backbreeding attempts to re-create extinct or nearly extinct species was the aurochs, a cattle species considered to be the wild ancestor of domestic cattle that became extinct in the 1600s. It had been widely distributed across Europe and North Africa in the Pleistocene, but by the Middle Ages, it existed only in small numbers in Central and Eastern Europe after declines due to hunting and habitat loss. The last aurochs died in 1627. Or did it? In the early 1920s, German brothers Heinz and Lutz Heck started breeding cattle to create an aurochs. The resulting animals, known as Heck cattle, have been used in rewilding projects to replace extinct wild cattle. In the late 1990s, Heck cattle were further bred to continue moving the phenotype toward the extinct aurochs. The Tauros Programme is pushing this attempt even further, focusing on genetically matching its tauros breed to extinct aurochs. In the Prehistory gallery of the Museum d'Histoire naturelle in Nîmes, France, one of these "new" aurochs was on display. Its label had a very personal touch: "My name is Hildegarde. I am a new aurochs! I resemble the ancestors of the cattle which lived during the prehistoric period up to the 16th century [sic]."

The quagga breeding follows the same trajectory as the aurochs. Sandra Swart has written about the history of the quagga's

de-extinction and explains that taxidermist Reinhold Rau, who worked to restore the South African Museum's quagga foal specimen in 1969, came to the conclusion that the quagga was simply a southern variation of the plains zebra. In 1985, he came together with a retired veterinarian to set up a project to breed zebras to make quaggas. The project, now under the name the Quagga Project, continues. Rau's inspiration to backbreed the quaggas had originally come from Lutz Heck's book on the aurochs he had helped make. The Quagga Project now labels its animals as "Rau's quagga," like Heck cattle, after the founder of the project.

Naturalis is not alone in presenting the quagga as once extinct but no longer. In the National Museum of Scotland, the quagga is placed in a case about recovery of threatened species. This particular animal, which died in 1872, is the only quagga that had been photographed while alive. A photograph pasted onto the glass in front of the specimen shows the quagga at the London Zoo in 1870. Like the Naturalis exhibition, this label also includes caveats about genes ("Despite the lack of stripes, quaggas are genetically identical to fully-striped East Africa zebras") and two paragraphs about de-extinction. Under the headline "Recreating species?," the text notes that because DNA degrades, genetic restoration of extinct animals is unlikely, but "in South Africa there have been recent attempts to breed back the extinct quagga from other zebras with reduced striping. This has been successful because the quagga's genes never became extinct." As in the Naturalis text, the quagga is presented as once extinct and as now existing through backbreeding. In both, the quagga as a unique species is deemed replaceable by a zebra bred to have fewer stripes.

The photograph of the London Zoo quagga is used in several other exhibitions as the only photograph of a living quagga, including at the Natural History Museum at Tring. The Tring exhibit hedges a bit more about the quagga's de-extinction, saying that "A zebra with quagga-like features may yet be seen again . . . The Quagga Project aims to selectively breed a population of plains zebras to look similar to the extinct subspecies."

Although the Scottish label implies that genetic de-extinction is unfeasible, there are other ongoing scientific projects to resurrect the thylacine, passenger pigeon, and more. The Pyrenean ibex (*Capra pyrenaica*, also known as the bucardo) was the first extinct animal to be born via advanced genetic techniques in 2003, though it only lived a few minutes. The de-extinction story is ignored in the display of the last bucardo in *El Museo del Bucardo*, where the exhibition does not mention the cloning even though the museum was opened in 2013, ten years after the short-lived de-extinction birth. Instead, the museum, which is located in

FIG. 45
The backside of the quagga without stripes, National Museum of Scotland, Edinburgh, UK, 2022.

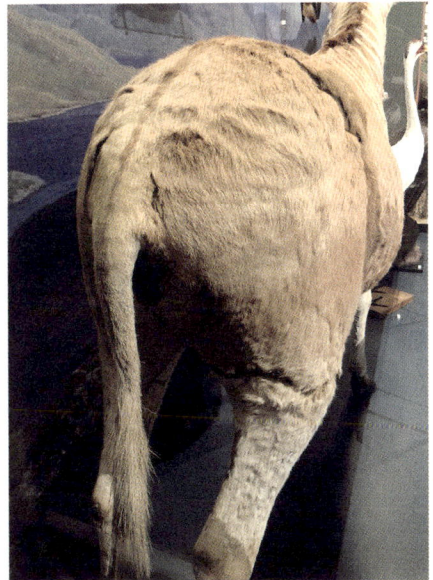

the mountainous region where the bucardo once lived, stresses the failure to conserve the bucardo and works to keep its memory alive. The de-extinction narrative was not relevant for the local organizers who wanted to honor the bucardo. As a contrast, in the Grant Museum of Zoology, in a context distanced from the local lives of the bucardo and the people of the Pyrenees, the comeback from extinction is the main feature of the display of the Pyrenean ibex.

The quagga's status as extinct and the Artis specimen's claim to be the endling are not questioned in all museums where it is displayed. In Paris, the Artis quagga's death in 1883 is mentioned as context for the museum's own quagga, which had also lived in a zoo—the royal menagerie at Versailles—before moving to the Jardin des Plantes (which later became known as the zoological gardens) in 1794. In Frankfurt, the label says

FIG. 46
Pyrenean ibex skull,
Grant Museum of
Zoology, London,
UK, 2015.

ADOPTED BY
The Celini Family

BACK FROM EXTINCTION
Using a goat as a host, a Pyrenean ibex
was the first species to be brought back
from extinction. It died after seven minutes.

that quagga were hunted for hides for grain sacks and the last quagga died in 1883 in the zoo in Amsterdam. In Berlin, the quagga is simply labeled as becoming extinct in 1883 because of intensive hunting. It is possible that the labels have not been updated since the quagga breeding projects, but it is also possible that they are unimportant to the narrative, just as the de-extinction was unremarked in *El Museo del Bucardo*.

There may be zebras with fewer stripes that come into the world, but are they really quaggas? It depends on how you define what a quagga is. Can a quagga be a quagga without being a particular type of being that lived and breathed in the world in its herd and passed on its way of being to its quagga foals? How will we know if the new quagga really has the distinctive bark, the "quácha" sound that it was named for, without recordings of old quaggas to compare against? As Swart asked, "Is there an organic quagga-ness and an indefinable identity predicated on millions of years of evolutionary history?" These are questions that are difficult, if not impossible, to answer.

Swart concluded her study of the quagga's history with "All animals are real, but some animals are more real than others." It's not a bad way to think of these quaggas. It doesn't mean the new quagga-like zebras aren't valuable animals themselves, or that they might not have a role to play in environmental conservation of the future, but I do think visitors encountered the endling quagga in that dimly lit treasures room in Leiden. The quagga are extinct, and I met the last of them.

6 *Broken Relations*

The thylacine pup looked so small and helpless. The large tag on its hind leg gave its specimen number (87.5.18.9), its species name (*Thylacinus cynocephalus* Harris), the location of collection (Tasmania), and the name of the collector (Tasmanian naturalist William Frederick Petterd). Towering above the little stitched and mended body was a picture of the last adult thylacine, which at the time was living in a zoo in Hobart, Tasmania. Underneath, a text explained that thylacine were intentionally eradicated because British settler farmers thought they threatened their sheep. Museum scientist Anjali Goswami offered her thoughts on the display: "To me, this baby Tasmanian tiger represents the fragility of the natural world. When we eradicated this species, we lost a very unique part of nature forever."

This display was part of the Natural History Museum's *Our Broken Planet* temporary exhibition in London. The exhibit

FIG. 47 Thylacine pup skin, Natural History Museum, London, UK, 2022.

covered myriad environmental pressures on the planet, including human-induced climate change, pollution, waste disposal, habitat loss, and extinction. It was easy to sense the broken planet in this little guy's broken body. A life cut short.

Extinction as a process doesn't just happen to one individual; it happens to populations. It breaks down herds and flocks. It means lonesome individuals cannot locate mates. It means unborn future generations, beings that will never come to be. Extinction severs relations in both time and space. The non-Western, Indigenous understanding of kinship is useful when thinking about these broken relations because kinship encompasses complex interactions between all kin (both humans and others) outside the biodiversity paradigm. Human–animal relations are sites of engagement and reciprocity, which can either help others thrive or cause them to die. Human-induced extinction, then, might be considered the ultimate failure of reciprocity—humans did not live up to our side of the bargain.

What kind of family relations are visible in extinction displays? What relations have been broken because of extinction? Thinking through museum displays of extinct relations—the eggs and nests, pairs and flocks of extinct species that inhabit museum halls—helps us consider the knots undone by extinction.

GIANT BROKEN EGGS

In the 1830s and 1840s, European travelers to Madagascar started collecting the huge eggs of the giant extinct bird *Aepyornis*, also known as the elephant bird, and bringing them back to European scientific collections. These are the largest eggs ever found. The first scientific publication about the bird, which was dubbed *Aepyornis maximus*, and its eggs appeared in 1850. The bird's mystique quickly invaded the public's imagination. In a literary work dated to 1851, Henry Morley, a professor of English literature at University College London, claimed that he wanted to be an *Aepyornis* to lay giant eggs of mischief, noting that "the fossil eggs of that bird now in Paris are sublime." According to Edward Hitchcock in *Outline of the Geology of the Globe* (1853), the previous existence of *Aepyornis maximus* had recently been made known through supersize eggs (as big as six ostrich eggs) and bones (the bird would have been ten to twelve feet tall) brought back from the island. The extinction of the *Aepyornis maximus* and its eggs were mentioned in popular science magazines such as *The Observer: An Illustrated Monthly Magazine of Interchange of Observations for All Students and Lovers of Nature* (1893) and *Natural Science: A Monthly Review of Scientific Progress* (1894). Illustrations of the fantastically large eggs appeared in popular magazines and books. They became an object of desire for European naturalists.

The science fiction writer H. G. Wells published a short story, "Æpyornis Island," in *Pall Mall Budget* in December 1894 premised on the *Aepyornis* egg hunt. In the tale, a sailor who collected island specimens for English gentlemen collectors had been collecting eggs and bones of the extinct *Aepyornis maximus* when he becomes marooned on a nearby small island. When the sailor ends up drifting at sea, he has three unbroken *Aepyornis* eggs with him. He cracks open the first one on the second day at sea and eats it (it tastes like duck). On the eighth day, he opens the second one, but it has a developing embryo. He thinks it is disgusting, but being stuck on a boat in the middle of the ocean makes a man hungry, so he eats it all anyway. The third is still untouched when he comes ashore on an island. The egg ends up hatching even though it was hundreds of years old. As it grows up, the bird becomes quite unruly and starts attacking the sailor. Growing more frustrated with the bird's aggressive behavior, the sailor ends up killing the bird and regrets it, feeling loss and loneliness. He saves the skeleton and sells it to a collector after he is rescued.

"Æpyornis Island" is fiction not only because the bird ends up hatching, but also because the collector finds so many intact eggs. The vast majority of *Aepyornis* eggs were broken because they had hatched, and collectors had to work carefully to put them back together. The egg on display in the Museo Civico di Storia Naturale di Milano is typical in this regard. All the pieces had to be joined together with glue. Spots with missing shell had to be filled in. The construction is held together in a frame. This gives the nominal appearance of a whole egg.

FIG. 48
Aepyornis maximus
egg, Museo Civico
di Storia Naturale
di Milano, Milan,
Italy, 2019.

The gigantic, flightless elephant birds became extinct on Madagascar soon after human settlement of the island. Although the exact extinction date is unknown, radiocarbon dating suggests the bird died out sometime around the year 1000–1200 due to human predation and disease transfer from domestic fowl. Eggs may have been an easily acquired food source for the human population. It is fitting, then, that there are so few intact eggs.

Elephant bird eggs are commonly displayed in natural history museums. As the largest egg in the world, it fits into the trope of "the wonders of nature," which has motivated natural history collections since their early days as cabinets of curiosity. The nearly unbelievable gigantic size of the eggs is the most prominent feature on labels, but size is a tricky thing to convey behind glass. The plain black backgrounds used in many displays make it difficult to judge the size, as there is little to immediately compare the egg against. Words have to do the work. Take, for example, the label at the Oxford University Museum of Natural History, which recounts a story to stress the egg's size: "Elephant bird eggs are the largest known and would hold the contents of 7 ostrich eggs or nearly 200 chicken eggs. One of the earliest descriptions, from 1848 by a French merchant, describes how the natives of Madagascar brought an empty egg onto his ship to be filled with rum for their chief which held 'the incredible quantity of 13 wine quart bottles of fluid!!!'"

The giant egg stands in for the giant bird. As the label in the Muséum national d'Histoire naturelle notes, "the egg exhibited here testifies to the size of this bird of which there are no stuffed specimens." Each egg is unique with its own patterns of light and dark, cracks and bumps, yet it has no individuality. It represents a bird that the viewer doesn't see. This "standing in" means that the egg is completely taken out of its context. It is not shown with a parent bird or in a nest. It is a broken egg in a broken world.

The only exception was a display of the *Aepyornis* as part of a special exhibit called *Mythic Creatures* at the Witte Museum in San Antonio, Texas. The premise of the exhibition was that some creatures of the imagination, such as dragons, are based on real animals. The *Aepyornis* in this context was discussed as the source of Marco Polo's roc, a bird so huge that it could lift elephants in the sky (the story is the source of the common name "elephant bird"). A model of the adult bird at full size stood next to a nest with a huge model egg. Although the text on the label was extremely similar to other elephant bird displays (many of them mention the roc connection too), this display had a very different feel about it. It showed a relationship between bird, egg, and habitat.

This display reconstructed a nest of the elephant bird, but its configuration is a guess because there are no existing nests. Nests of extinct species are much rarer than eggs in collections, and they hardly ever appear on display. The world's only remaining huia nest was on display at the Canterbury Museum in its *House of Treasures* exhibit in 2020, but the pandemic meant few outside the immediate area could visit it. I only saw its picture online. I was lucky enough to encounter a passenger pigeon nest in the special exhibit *Animalia: Animals in the Archives* at the Archives of Ontario in Toronto. The nest made of fine aspen twigs had been collected in 1883 in Manitoba; a model of an egg was added for effect. The nest was small, but proportional to the taxidermied passenger pigeons that were also in the case. The neatness of the twigs—an interlocking of chaos into pattern—was mesmerizing. Perhaps the collector had taken the nest with its original unhatched egg in 1883 for his oology collection unaware of the species being on the brink of extinction, or perhaps he collected it because there were no more pigeons to return to it.

FIG. 49 (*facing*)
Model of adult
elephant bird with
egg in nest, Witte
Museum, San Antonio,
Texas, USA, 2019.

There are a few displays that place other extinct bird eggs into relations. For example, great auk eggs were put together with an adult bird in Copenhagen's Statens Naturhistoriske Museum and in the Naturalis Treasures exhibit. The illusion is that

FIG. 50
Real passenger pigeon
nest with a model egg,
Archives of Ontario,
Toronto, Canada, 2019.

the egg and the adult belong together, which of course they do not—many times they were collected years, even decades, apart. But illusions to heal the broken are integral to natural history museums.

A STRIKING PAIR

Huias (*Heteralocha acutirostris*) are almost always displayed in pairs. This extinct wattle bird native to Aotearoa New Zealand has striking sexual dimorphism, which means that the males and females have a radically different appearance. In the case of huias, the beak of the female is long, thin, and curved downward, whereas the male's is short and stubby. This was a major draw for nineteenth-century naturalists who wanted to add both to their collection.

The huia became extinct around 1907 due to extensive

hunting. It had been under pressure since colonization of the island by Māori, who used the white-tipped black tail feathers as marks of high status. Feathers of this sacred bird (*taonga*) were worn in the hair, and skins could be dried and worn as ornaments. The coming of European settlers quickened its decline, as they hunted the bird both for museum collections and for the feathers that were prized as hat trimmings. Taxidermist Andreas Reischek, for example, sent eight huia among a nearly complete collection of the birds of New Zealand to the new natural history museum in Vienna. In 1892, the government extended the Wild Bird Protection Act to include huia, but it was too late to save the bird from extinction.

Like their divergent beaks, huia pairs sang to each other with very different songs. In 1948, long after the last huia died, a Māori elder, Hēnare Hāmana, sang the calls of the huia at the request of local historian Robert Batley. Hāmana had been a huia mimic involved in searches for the bird in about 1909 to find individuals for conservation. The call of the male included fast, repeating notes, whereas the female had only a smooth movement between three notes. They were in conversation as a pair, so it is fitting that they are displayed as pairs in many museums. But while the huia displayed in pairs might look to be unbroken families, in reality the specimens on displays were rarely a pair in life.

There are a few natural history museums with just one huia on display. In this case, the bird chosen is always the female with its long and elegant curved beak. This is an aesthetic choice that runs counter to what we find in most natural history museum displays. More often than not, male individuals are put on display because of their showy features, such as large horns or flashy, brightly colored feathers. As Donna Haraway

FIG. 51
Huia pair showing
their different beaks
and the female in
the higher position,
Musée d'Histoire
naturelle de Lille,
Lille, France, 2022.

has noted, naturalist hunter-collectors like Carl Akeley of the American Museum of Natural History preferred adult male specimens as quarry. Rebecca Machin studied the public displays of the Manchester Museum in 2008 and found that 71 percent of the mammals and 66 percent of the birds in the cases were male. In the bird gallery, 48 percent of the species had both genders on display but another 44 percent had only males. The view is that female birds and mammals "are the browner, smaller, duller, uninteresting and unimportant variants of the males that are preferred for display."

In addition, Machin found that males are often placed in higher positions and more erect postures than females when they are displayed together to indicate male dominance. But the huia goes against this trend. Since the huia female is the one with the impressive beak, it is typically placed higher than the male. When only one animal is displayed, it is the female that is chosen. Yet even this choice reveals that natural history museums are following the same patterns for the huia as they do for other displays, picking the most aesthetically bold specimen to highlight.

There is another broken relationship that the visitors in the museum will not see: the huia and its parasite. *Rallicola extinctus*, which was described after its extinction and named accordingly, was a host-specific feather louse of the huia. The louse was described in 1990 by Eberhard Mey based on adults, larvae, and eggs collected from museum

FIG. 52
Lonesome female huia, Kelvingrove Art Gallery and Museum, Glasgow, UK, 2022.

storage specimens in Germany. Mey originally named the louse *Huiacola extinctus*, but the name was later changed to reflect the genus to which it belongs. Mey argues that based on the morphological comparison with other feather lice, it must have been coevolving with huia for a long time. A specimen had also been collected from huia before their extinction: one in the Te Papa collection in New Zealand was collected in 1880 from a huia, but it was not identified until after the Mey publication.

Species-specific parasites like *Rallicola extinctus* go extinct when their hosts do. We know of extinct parasites associated with other extinct birds, including the passenger pigeon (two different lice), Guadalupe caracara, New Caledonian rail, Guadalupe storm petrel (one chewing louse and one louse), Jamaican petrel, and Carolina parakeet (six feather mite species). Measures are sometimes taken in modern conservation projects to avoid causing the extinction of species-specific parasites, but it is too late for these. While specimens of extinct lice like *Rallicola extinctus* sit in the back-room collections, these extinctions are not included in museum narratives. The relations are broken.

MIGRATING TO THE MUSEUM

The Labrador duck (*Camptorhynchus labradorius*) is named after its assumed breeding grounds in Labrador, Canada, on the far northeastern edge of North America. When John James Audubon described them as the "pied duck" in the 1840s, he reported that their nests occurred in Labrador and were similar in size and construction to eider duck nests. Little evidence beyond Audubon's report exists about the nesting of the birds, so he might have been wrong about which nests belonged to which birds. He called the pied duck "a very hardy bird" that could be seen along the coasts from Nova Scotia to the Chesapeake Bay even in severe winters. He illustrated the birds for *Birds of America* from a pair killed by the statesman Daniel Webster on Martha's Vineyard, Massachusetts. That pair made their way into the Smithsonian's collection. It wasn't long after Audubon drew them that the species was gone. Hunting pressure, particularly egg collection for consumption, like with many other ducks, likely led to its eradication. The last confirmed spotting of a live individual was in its wintering ground in New York in 1878; it was eaten for dinner.

The Labrador ducks in the American Museum of Natural History in New York are shown as a small flock wintering at Montauk Point, Long Island. The four taxidermied individuals (one male and three females) are in a snowy habitat diorama with a painted backdrop showing more ducks nearby. The ducks have been placed into their wintering grounds in New York, where the museum is, rather than their summer nesting grounds in the north. The makers of a diorama always have choices to make about the setting that they put animals into. The choice makes the females' brown bodies pop out of the landscape as a contrast to the stark white snow. Because the ducks migrate, the choice also changes their geographical association. While they are named for Labrador, the diorama places them in New York.

A similar move with a migratory flock happens in the "Hall of New York City Birds" of the same museum. This time it is the passenger pigeons. A diorama shows a flock of passenger pigeons and a pin oak tree in autumn in New York. The group is lively with both males and females, some on branches, others on the forest floor collecting nuts. The label includes the

FIG. 53
Labrador duck diorama
set in Long Island,
New York, in the duck's
winter feeding grounds,
American Museum of
Natural History, New
York, USA, 2022.

nugget that "they were sold for as little as a penny apiece in New York markets and were even fed to pigs." This display gives the passenger pigeon a relationship with New York, where it is now viewed, rather than farther west where most of the birds were hunted.

There is an even larger diorama of passenger pigeons in a woodland in the Academy of Natural Sciences in Philadelphia. The museum, which is the oldest natural history research institution in the United States, having been founded in 1812, has an incredible collection of dioramas displaying the ecological diversity of North America, Asia, and Africa over several floors. On the third floor, passenger pigeons are tucked into the branches and pecking through fallen leaves on the forest floor of a large in-wall diorama. I am not sure how many pigeons were in the display, as I lost count around fifteen, unsure of which I had already counted. There is an irony of not being able to count the pigeons in the diorama. Pennsylvania had seen flocks of millions of passenger pigeons in the 1800s. A monument commemorating their loss was set up in October 1947 by the Conewago District Boy Scouts in an area known as Pigeon Hills near Harrisburg, where the passenger pigeon had once darkened the skies when they migrated overhead.

The diorama, standing where there had been woodlands and passenger pigeons, gives a glimpse of both now long gone. The artificial sunlight illuminated the clearing among the deciduous trees where the birds and leaves blend together. The scene invites the viewer to push the branches back and step in to hear the chatter in the wilderness. But, of course, that is impossible, as the birds stand frozen behind glass.

Both the Labrador duck and the passenger pigeon displays use their flocks to display multiple individuals in different natural positions. The male duck is moving his wing over his head; a female pigeon near the top of the branches stretches her wings out to take off. The birds are seemingly gesturing to one another as they walk or forage. It gives the viewer an impression of the movement of a flock, but without the sound that surely must have been loud, in the case of the pigeons at least.

There is a second diorama near the passenger pigeons in the Academy of Natural Sciences, but it offers a view that never

FIG. 54
Diorama of
passenger pigeons
in a Pennsylvanian
woodland, Academy
of Natural Sciences,
Philadelphia, USA,
2023.

FIG. 55
Diorama of three
extinct northern water
birds that did not
live in the same area,
Academy of Natural
Sciences, Philadelphia,
USA, 2023.

could have been. Two Labrador ducks stand on a rocky cliff in front of the sea—the white-headed male high up on a ledge with the duller-brown female on another ledge closer down by the viewer. Joining them in the scene is a great auk standing proud in the middle and two Eskimo curlews, one on a ledge opposite the male duck and the other hidden on the cliff's edge by the sea. These birds never would have been in the same scene in life, although the label notes that they had "something in common," that is, that they are water birds from the north and, "sadly, they are all extinct." This particular diorama misleads the viewer into thinking they are seeing a glimpse of a lost ecosystem. It places the Labrador ducks, great auks, and Eskimo curlews into relations with each other, but these are connections that apply only in death, not in life.

These dioramas give a sense of relations among the birds and between birds and place, but it is a necessarily limited view of what has been lost with their extinctions. The constraints of the medium mean that only one environment at one time of year can be displayed. Individuals who never lived together, and may have in fact lived and died decades apart, inhabit the same physical space. All the broken relationships can never be captured in the museum.

THE MISSING YOUNG

While natural history museums commonly display eggs and adults, even in pairs or groups, the young are virtually absent. That was why the young thylacine in *Our Broken Planet* that started this chapter was so striking—it is very rare to see young extinct animals on display.

But there were a few other young thylacines on display in the past. There was once a taxidermy group of a thylacine mother lying on her side with four small pups, one tucking itself into her pouch. Kathryn Medlock traced the taxidermy preparation, which was given to the Tasmanian Museum in 1884 by the Buckland and Spring Bay Tiger and Eagle Extermination Society. The Society was founded that year to pay farmers a reward for killing thylacines, so it had bodies to spare. While it may seem strange that a society that wanted to exterminate a species also wanted to display the animal, it has to be set into the Victorian context of museums for public education. They wanted the carnivore gone at a practical level, but that didn't mean they didn't appreciate the thylacine as part of nature at an intellectual level. The female with young was hailed by a newspaper as an exhibit of "graphic ingenuity," "one of which any museum might well be proud." The exhibit was destroyed in 1935 as part of a clean-out of the museum stores—an extinction of those young thylacines from future view.

The other was a thylacine pouch pup on display as a wet preparation in a jar at the Australian Museum. This pup, which had not yet gotten big enough to leave its mother's marsupial pouch, was preserved in 1866 in alcohol rather than formalin. That means that its DNA is still at least partially intact. DNA was extracted from the pup as a first step in the Australian Museum's Thylacine Project (1999–2005), which had the idea to bring back the thylacine using genetic cloning technology. The project was heavily criticized in the media. Although the pup itself was heralded as part of the de-extinction story in the museum presentation, according to one former employee, it yielded almost no useable DNA. This specimen was not on display to show the horrors of extinction, but to show the hopeful possibility of reversing it. That is an understandable storyline given the museum's involvement, but it is somehow unsatisfying to imagine only positive futures given the negative past. That's where the *Broken Planet* juvenile was different. It was there as a poignant reminder of the missing children, not a sign of future children.

There are other young still on display that can remind us of the broken families now extinct. One is a Formosan clouded leopard cub at the National Taiwan Museum. The small blond body with dark-brown cloudlike spots stands on a glass plate inside a case of animals named for Taiwanese geography. The Formosan clouded leopard (*Neofelis nebulosa brachyura*)—Formosa was the English name for Taiwan—was scientifically described in 1862 by Englishman Robert Swinhoe, the first Westerner to conduct systematic collecting of Taiwanese fauna. It did not take long for deforestation and industrialization in the wake of

FIGS. 56 & 57
(*facing*) Formosan clouded leopard cub and adult in two separate exhibitions, National Taiwan Museum, Taipei, Taiwan, 2023.

過去的未來

The Past Is the Future

Taiwanese colonization by Japan and then modernization after World War II to make the island uninhabitable for the leopard. This top forest predator was declared extinct in 2014 after extensive forest surveys found none. The little leopard seems helpless. It is old enough to have its eyes open, but certainly not old enough to be on its own. Yet on its own it is.

An adult Formosan clouded leopard is on display on the floor above, in the last room of the *Discovering Taiwan* exhibition, which chronicles the history of natural history on the island. Here, like the Australian Museum thylacine pup, the narrative of the leopard centers on the possible future for the extinct species. The adult stands with its long tail curving upward in a dramatically lit case under a colorful map of the island. Under the heading "Vanishing and Rebirth," a nearby information board notes that "museum collections might provide unexpected functions in later years," such as measuring DDT concentrations in eggshells and collecting passenger pigeon DNA. This is the same kind of narrative constructed in the Australian Museum for the thylacine project. The museum's leopard remains "contain important genetic and morphological information that might lead us to revive the lost in the future," according to the label. It seems that the promise of a future with live leopards is more attractive than the present haunted by the cub on the second floor.

Another young specimen is a little quagga on display in the Naturhistoriska riksmuseet in Stockholm. It has big wide (glass) eyes, with the clear reddish-brown and white stripes. The skin looks a little too tight on its face, and many of its seams are breaking open. The label says this is "Sparrman's quagga." It is one of the museum's oldest objects, brought from South Africa by Anders Sparrman, a student of Linnaeus and first curator of the collection of the Royal Swedish Academy of Sciences, in 1775. The label also says this isn't a foal, but a fetus

FIG. 58
Sparrman's quagga,
which was collected
as a prenatal fetus,
Naturhistoriska
riksmuseet, Stockholm,
Sweden, 2021.

(*fostre* in Swedish). It was killed before it was even born to become a collector's object. "Because the species is now extinct, the quagga foetus acquired in 1775 is one of the museum's most valuable objects," the label tells us. Here is one of the missing children of the extinct.

We are left with broken relations in extinction exhibitions. We can think of the little ones that were killed as unhatched eggs, nestlings, or pouch pups that don't make appearances in public. We can think of the pairs that will never sing together and the flocks that will never fly. It reminds us of the family lives of the animals that were broken and cannot be repaired.

7

Memorialized Dead

The cell-like wall of windows and metal bars towers above you. Light streaking in through the Shanghai Natural History Museum's organic-looking facade gives you a sense of being a small part of a larger web of life. A purple timeline runs the length of the second-floor interior wall opposite the windows. Along the timeline are huge images of animals and plants that have become extinct in the last 400 years along with the date of their demise, starting with the aurochs in 1627 and ending with the baiji (Yangtze River dolphin) in 2002. The display, titled "Memories of Life," shows "fading figures" that have "left us sad memories," according to the label. The form of this memory-making in Shanghai was monumental and collective. It was not a statue to one animal, nor a small display. It was a larger-than-life roll call of the dead.

Monuments are commonplace globally for the remembrance of the dead, from grave headstones to war monuments.

1700 1680 1627

1600 1500 1450 1400

These practices include animals as well as humans, from pet cemeteries to warhorse memorials. Whereas memorials to public figures and major events have a long history, the US Civil War was a turning point in the commemoration with a new type of monument to the ordinary soldiers who died. This type of memorial that lists the fallen is most well known today in the Vietnam Veterans Memorial designed by Maya Lin and dedicated in 1982. That black granite memorial rises and falls across a Washington, DC, plaza, engraved with the names of casualties, that comes to overshadow the visitor before it fades to nothing. Outdoor monuments for extinct species, from stone cairns to sculptures, have been put up in this same vein. Stone monuments set up for the extinct passenger pigeon in the post-WWII period show that they were thought of as "casualties of war, although it was a war on nature, and were understood at the time as worthy as objects of grief."

What is the function of memorializing the extinct in museums? The Shanghai Museum's timeline offers to fix the fading memories on the wall to help visitors remember, even though the memory will be sad and painful. In this extinction timeline, we find two elements—the name and the image—each of which are explored in this chapter in the context of memorials and memory of the extinct in museums. These two elements in museums are one potential avenue for remembering the dead.

ENGRAVED NAMES

Naming is a key feature of the human relationship with non-humans. Local human communities name animals and mountains, streams and plants as part of ordering the world. The biblical story of God bringing the animals before Adam to name them (Genesis 2:19) is not really about control, but about relations. To be in personal relation, you need to know someone's name. As Indigenous scientist Robin Wall Kimmerer observes, "Names are the way we humans build relationship, not only with each other but with the living world. I'm trying to imagine what it would be like going through life not knowing the names of the plants and animals around you. . . . I think it would be a little scary and disorienting." In his study of naming as a route to anticipating the loss of extinction, Joshua Trey Barnett sums up the value of the name: "naming, simply put, draws us nearer to others in life and in death."

Modern scientists give names to species through the Latin binominal system. This system, which originated with the naturalist Carl Linnaeus, gives two names to a species: the first part of the name is a genus (roughly a "kind"), and the second part is an individual identifier within the genus. This (in theory, although not always in practice) groups species together by how similar they are to each other. Linnaeus developed his system prior to Darwin's theory of evolution, so the naming schemes are not strictly evolutionary, yet things that are evolutionarily related are often in the same genus, such as horses, zebras, and donkeys being named as *Equus something*.

Both the local and scientific flavor of names matter. The scientific name allows portability across languages, so that it is always *Thylacinus cynocephalus* whether it is labeled in a museum as thylacine (English and French), buidelwolf (Dutch), beutelwolf (German), pungulv (Norwegian and Danish), Wilk workowaty (Polish), Lobo marsupial (Spanish), or フクロオオカミ (Japanese). Binominal names help scientists know that they are talking about the same animal. Colloquial local names are how people talk about species on a day-to-day basis. Those names have specific historical roots and implications—like the names of *Thylacinus cynocephalus* in many languages referring to it with a word meaning "wolf." There has historically been an erasure of Indigenous names of species. The Aboriginal people of Australia called the thylacine by different names, such as *coorinna* and *laoonana* in various languages. But those don't appear in any thylacine exhibit, even in Australia.

What *was* in Australia was an aluminum metal box illuminated by multicolored lights: the "Endling" display within the *Old New Gallery*. The endling implied was the thylacine, the last of which died in 1936, although the skin on display was not from that thylacine but rather from one that was hunted in 1930. The inside of the box is less pertinent than the outside for this discussion of monuments because on the outside were engraved names. The engraving of the names made it feel monumental and permanent (although, like all museum galleries, it is in fact not permanent and was removed in 2022 and replaced by a new thylacine exhibit in the gallery *Great Southern Land*). The names were Australian species that had gone extinct since the British colonizers arrived in 1788. There was a mausoleum feel about the box, a tomb for the extinct thylacine covered in names of the dead family.

Pig-footed bandicoot (*Chaeropus ecaudatus*). Eastern hare-wallaby (*Lagorchestes leporides*). Percy Island flying fox (*Pteropus bruneus*). More and more names could be seen around the box on all sides. All had the English settler common name and a scientific name, except most of the plants, which had only a scientific name. The choice to include only the settler name and not the Aboriginal one(s) was a practical one (there are many Indigenous languages in Australia, so which would be selected?), but it also downplayed the ongoing relationship of the

Aboriginal people to the species that have disappeared. Louise Boscacci has discussed how a museum specimen of a bilby elicited a response of love and affection from an Aboriginal elder who hadn't seen one since her teenage years: the elder Alice cradled the bilby (she called it *ninu*) skin, sang to it, and shared stories of the bilby's way of being. The Aboriginal population has names and stories for these animals, but these were not recorded in the exhibition. People and animals had coexisted for thousands of years, but now the relationship was broken.

In spite of the Aboriginal name shortcoming, the Endling box bore witness to the fallen by giving them names. This is a strategy that has been adopted at other museums as well. At the National Museum of Scotland, the wall next to the extinct animal case in the "Survival" gallery has a massive list of names overlaid on a photograph that extends to cover the long wall. The photograph is a famous image of a pile of plains bison (also known as buffalo) skulls at the Michigan Carbon Works in Rougeville taken in 1892. Two men in suits stand with the pile, a visual record of the capitalist market for buffalo bone, which was rendered by the Carbon Works into fertilizer, glue, and ash. The image evokes the rapaciousness of North American settlement and the destruction of nature, as well as the destruction of the way of life of the Indigenous tribes of the plains that depended on the bison.

The list of names has the title "In Memoriam 1600–2020," offering no doubt as to the memorial nature of the display. Below are the common English names for extinct species, along with a year, beginning with the large sloth lemur in 1620 and ending with the Bramble Cay mosaic-tailed rat in 2016. Some of the names are in bold font—these are species that the museum has on display in the case to the left. Some of the honored

dead in this museum are rare to see exhibited. This is the only museum I visited where visitors can encounter the Ascension Island crake, bush wren, Curio's giant rice rat, Guadalupe storm petrel, New Zealand quail, and Polynesian tree snail *Partula turgida*.

The American Museum of Natural History had a similar list on a wall in its biodiversity exhibition in 2015. A box with white lettering on a red background at the top of the black panel announced, "Not since the end of the age of dinosaurs—the great extinction at the end of the Mesozoic era—have species disappeared at such a rapid rate." Below and to the sides in a small font (to fit in all the names), the list gave the scientific name and English common name for species extinct since 1500 ordered by type (mollusks, crustaceans, insects, birds, etc.). The panel, however, noted that the list is incomplete. In the center section, additional text in a large font warned the visitor that "it is certain that many more species have become extinct in the past than are listed here: they have lived and died without our ever having learned of them." This was a way of invoking the unnamed, unknown in the tomb.

The Museum der Natur Hamburg decided to use a list of extinct species as part of its new exhibition case on the Anthropocene, "Anthropozän: Das Zeitalter des Menschen." A long strip of paper showing the German common name, scientific name, date of extinction, and reason for extinction extended from the top to bottom of the case with numerous folds. The display choice emphasized the length of the list. It showed both worldwide extinctions and local extinctions within the German national borders. Including the German common names for the entire list rather than just for the German species made the animals more recognizable to the German audience. The museum had no organic remains of extinct species on display, but did use visual images of extinct animals in the video in the Anthropocene case. This video turned the case contents, including the extinction list, into a more dynamic display. The list of species scrolled over a background image of various extinct animals, including an eastern hare-wallaby, quagga, and thylacine.

All the exhibitions above emphasized the abundance of extinct species. The long lists of names are intended to convey the scale of the loss. Instead of just using numbers to signify total extinctions, these exhibitions used names to call out individual species losses. Naming the lost—perhaps engraving them on the hearts of visitors—is a step toward knowing them.

PUTTING FACES TO NAMES

John James Audubon, a French immigrant to America, decided in about 1820 that he wanted to create a painted catalog of every bird in North America. Rather than just producing another book of words, he wanted to produce images to reveal nature's wonders. He ended up publishing 435 hand-colored life-size prints of 497 bird species in his series *Birds of America,* which was printed in a series of installments between 1827 and 1838. Audubon was not the first to try to record all the birds of North America—Alexander Wilson had previously attempted it with *American Ornithology or, the Natural History of the Birds of the United States* a decade before—but never had the birds been depicted at life size. Audubon chose to depict his birds on sheets measuring about forty by twenty-nine inches with a plant species relevant for the species. The birds are often depicted interacting with other birds or their own prey. Audubon and ornithologist William MacGillivray produced a separate accompanying text, *Ornithological Biography, or, An account of the habits of the birds of the United States of America,* in five volumes between 1831 and 1839. While there were later printings of *Birds of America* in smaller formats, it is the original "Double Elephant Folio" size that is highly attractive for exhibition. Because the prints were issued loose in tins and then later bound by subscribers, these large folios are often available for display on the wall rather than in book form.

Audubon's bird illustrations are works of art as well as works of science. While Audubon has a controversial racial legacy as a slaveholder opposed to abolition and a looter of Native American grave sites and human remains, it is no wonder that his illustrations are highly desirable for exhibitions. A number of Audubon's birds are now extinct: the Carolina parakeet (labeled as Carolina parrot), great auk, heath hen (labeled as pinnated grouse), ivory-billed woodpecker, Labrador duck (labeled as pied duck), and passenger pigeon. The Eskimo curlew and Bachman's warbler are classified by IUCN as "critically endangered, possibly extinct" and have not been seen for half a century (the US Fish and Wildlife Service decided in 2023 to declare the Bachman's warbler extinct). Although there are 435 prints to choose from when exhibiting *The Birds of America*, extinct birds often take center stage.

Teylers Museum in Haarlem, the Netherlands, put on a special exhibition *Vogelpracht* in 2021–2022 that featured Audubon's work. Teylers was one of the few European continental subscribers to *The Birds of America* when it originally came out. The museum bound the plates into five volumes (as opposed to the typical four). In order to show more images than just four in the open volumes without unbinding them, facsimiles from Teylers's exemplars were hung on the wall. The wall featured reproductions of *The Birds of America* plates for five extinct species (Carolina parakeet, great auk, heath hen, ivory-billed woodpecker, and passenger pigeon). The interpretive sign titled "Changes in Bird Paradise" next to these extinct birds offered Audubon's images to "highlight the differences compared to today's observations—sometimes to a painful extent." The text noted that Audubon "was one of the first to express concerns about the bird population," although he couldn't imagine populous birds like the passenger pigeon going extinct.

Even though the Carolina parakeet (*Conuropsis carolinensis*) drawing was included in the set of portraits of extinct birds, it also took center stage in the exhibition. In the room facing the visitor when you walked in, Audubon's illustration of the Carolina parakeet had been blown up to fill the wall with a taxidermied parakeet tucked into the image in a glass inset. Bringing together the drawing and physical body reminded the visitors that these birds were not just figments of artistic imagination. The drawings perfectly mirrored the brightly colored specimen with its yellow head, red face, and green body.

Audubon's Carolina parakeet illustration is a masterful work. Seven parakeets are shown as a flock on a common cocklebur. Both sexes of the parakeets have dark green feathers on top of their bodies and brighter green on their stomachs. Their heads are bright yellow with a red mask from eyes to beak. The exception is the young parakeet in the lower middle, which is still only green—the fully colored plumage is not acquired the first year. The birds are an overlapping tangle of feathers and claws that ends up showing the observer the bird from all angles.

The birds have alighted on a cocklebur, an annual plant that grows about four feet high and produces fruits covered in hooked spines. There can be thousands of the cocklebur fruits on a single plant. Audubon describes the parakeet's movements to open the fruit as a highly choreographed move: it "alights upon it, plucks the burr from the stem with its bill, takes it from the latter with one foot, in which it turns it over until the joint is properly placed to meet the attacks of the bill, when it bursts open, takes out the fruit, and allows the shell to drop." The image shows the parakeets in this act.

All the beauty in this supersize image obscures the

intentional destruction of the parakeet. In *Ornithological Biography*, Audubon describes the Carolina parakeet as "always an unwelcome visitor to the planter, the farmer, or the gardener" because it destroys every kind of fruit and grain. "Nature seems to have implanted in these birds a propensity to destroy, in consequence of which they cut to atoms pieces of wood, books, and, in short, every thing that comes in their way." The farmer retaliates with guns while the birds attack grain sacks. Audubon notes the efficacy of the counterattack; "I have seen several hundreds destroyed in this manner in the course of a few hours, and have procured a basketful of these birds at a few shots, in order to make choice of good specimens for drawing the figures by which this species is represented in the plate now under your consideration." In addition, the birds, especially the young ones, were hunted as food. This destruction was taking its toll. Audubon remarks that "Our Parakeets are very rapidly diminishing in number; and in some districts, where twenty-five years ago they were plentiful, scarcely any are now to be seen."

It took less than a century after Audubon's comments for the Carolina parakeet to become extinct. The last one, a male called Incas, died in 1918 in the same aviary in the Cincinnati zoo where the last passenger pigeon had died four years earlier. Unlike the 100th-year anniversary of the passenger pigeon's passing in 2014, with dedicated exhibitions at museums across North America and several popular science monographs gracing the bookstore shelves, few seem to have marked the Carolina parakeet's centenary. There were no special exhibitions to mark the loss of the brightly colored bird. It is not clear why these didn't happen, since many museums have Carolina parakeets as taxidermy or study specimens and they are striking colors, but it may have had to do with the closeness in time to the passenger pigeon centenary that had attracted so much attention and funding. It was heartening to see the bird take central place in *Vogelpracht* and get some attention.

When the National Museum of Scotland developed its own exhibition, *Audubon's Birds of America*, it too highlighted the Carolina parakeet. The parakeet was the key image for the exhibition marketing materials, and the green used as the main color of the exhibition was chosen to match the feathers in Audubon's drawing. This extensive exhibit reviewed Audubon's career and the making of the famous drawings, as well as his ties to Edinburgh as a publishing location for much of his work. The museum displayed numerous taxidermied specimens of the birds shown in the drawings, despite Audubon's insistence that "I have never drawn from a stuffed specimen" (this quote appeared on one of the gallery walls). The parakeet drawing features in a video available on demand about Audubon's skill at depicting lively birds, in contrast to the accepted scientific standards at the time.

The final section of the exhibition was titled "A Fragile Beauty." Here, Audubon was framed as one of the first to comment on species decline, and the exhibit noted his legacy as the namesake of many conservation associations but also his failure to realize the reality of extinction. An extra-size print of the Carolina parakeets, just like at Teylers, was put on the right-hand wall. The original drawing at the correct size was hung on the opposite wall with a label quoting Audubon's observations on the species' decline. A Carolina parakeet stood in a case in front of the large-scale print. The specimen's label noted that the birds were killed as farm pests and hunted for plumage; the last confirmed sighting in the wild was in 1910, eight years before the last captive bird died. But the object in the case was not a taxidermied Carolina parakeet—this was instead a faux parakeet created from other feathers to mimic a parakeet. The museum decided that its specimen was too fragile for exhibition (and the traveling exhibition that would follow after the Edinburgh showing), so a replacement was created. Although this information was shared on the museum's website, the exhibition itself simply labeled it as "Taxidermy copy of a Carolina parakeet," so it may not have been instantly clear that the thing in the case was not the remains of a Carolina parakeet. Using a model like this is not unusual (indeed, all feathered dodos are models made with other birds' feathers), but the museum could have used the extinction (thus irreplaceability) of the parakeet more explicitly to present why the museum's own Carolina parakeet was not in the room.

Next to the Carolina parakeet was a separate cabinet on the use of feathers in nineteenth-century fashion and the extinction of the passenger pigeon. Unlike the Carolina parakeet, the passenger pigeon in this vitrine was a taxidermied specimen. It was displayed next to a double-barreled flintlock fowling gun so there would be no doubt about why it became extinct. Other extinct birds made an appearance in the exhibition as well. The museum's copy of Mark Catesby's *The Natural History of Carolina Florida and the Bahama Islands*, volume 1 from 1731 (the first illustrated scientific description of America's birds), was open to an ivory-billed woodpecker.

The final room of the special exhibition showed a looping film on extinction, pointing out the losses of birds since Audubon and asking, "What does the future look like for birds, 80 years from now?" Even though Audubon himself was not concerned with extinction, the museum curators wanted to stress

FIG. 61
Carolina parakeet
faux taxidermy and
gigantic reproduction
of Audubon's plate,
National Museum of
Scotland, Edinburgh,
UK, 2022.

FIG. 62
Ivory-billed woodpecker
illustration in Mark
Catesby, *The Natural
History of Carolina
Florida and the Bahama
Islands*, vol. 1 (1731),
National Museum of
Scotland, Edinburgh,
UK, 2022.

that visitors today should be. Audubon's cataloging project takes on resonances that he would not have expected.

Audubon's drawings always featured the scientific Latin binomial name and the common name for the species (many of these names have been updated since his publication) at the bottom. This information gave it scientific credibility and international portability, but it isn't what attracts people to see them. Knowing a name is one thing, but knowing a face is something else. The art of Audubon creates an encounter with an eyewitness to the birds—Audubon always drew his birds from recent specimens, and he encountered many (although not all) in the wild. Although Carolina parakeets died after photography was invented, there are remarkably few images of the birds: one is a photo taken by Smithsonian curator Robert Shufeldt (who dissected Martha the endling passenger pigeon to determine cause of death), and the other is a snap of a Carolina parakeet kept as a pet by another Smithsonian scientist. Neither of these animals lived in the wild. Audubon's drawings are the best thing we have for encountering the Carolina parakeet as it would have been in life.

REMEMBERING THE DEAD

It is fitting that the extinction timeline memorial in the Shanghai Natural History Museum ends with the baiji (*Lipotes vexillifer*). It was a freshwater river dolphin living in the Yangtze River system of China. As the Yangtze River was heavily industrialized for fishing, transportation, and hydroelectricity, including the creation of the massive Three Gorges Dam that started to be built in 1994, the dolphin population plummeted quickly. Already in the 1980s, scientists knew the baiji was in trouble. Ex situ conservation efforts at dolphinariums were set up, but no baiji reproduced. The Chinese government approved a Conservation Action Plan for Cetaceans of the Yangtze River in 2001, but it was too late. The last confirmed baiji died in 2002. After a six-week survey of the Yangtze in 2006 in which no baiji were located, scientists concluded that "the baiji is now likely to be extinct."

The baiji is the only modern extinct specimen that the Shanghai Natural History Museum has on display. The skeleton is displayed in front of a text-filled backdrop. The outline of the baiji's fleshy body is created by changing the color of the background and text, animating the lifeless skeleton. The text repeats over and over again in imperfect English: "Various animals once lived in Shanghai. Some animals, such as

the Yangtze River dolphins, the badgers, roe deer, Whooper Swans, and Pallas's Sea Eagles, are barely found in Shanghai, even have totally disappeared. Their leaving is the regression induced by the rapidly expanding modern cities, sounding to us warnings of a changing ecosystem."

Even the animals on this list do not capture the biological losses in the Yangtze. The Chinese paddlefish (*Psephurus gladius*) was not on the museum's timeline because its extinction has been acknowledged too recently: it was declared extinct on the IUCN Red List in 2022, after the specialist group decided in 2019 that it should move from critically endangered to extinct. It was last seen alive in 2003. *Our Broken Planet* at the Natural History Museum in London had one on display. The giant animal was one of the largest freshwater fish on Earth, reaching over ten feet in length, but its habitat was irrevocably damaged by the Yangtze hydropower development project. "Renewable energy can threaten wildlife too," proclaimed the label with the paddlefish. The text made the point that even though hydroelectric dams are needed for renewable energy, there are "casualties of the global boom in hydroelectric dams." In the jar, which towered up and yet somehow seemed too small for the massive grumpy fish, it indeed looked like a casualty of a war on nature.

Shanghai's "Memories of Life," Scotland's "In Memoriam" plaque, Hamburg's printed list, and Audubon's *Birds of America* are all attempts to capture the vastness of life on Earth. The exhibitions are overwhelming. The lists and images go on and on. While there is value in communicating abundance—or, in this case, the loss of that abundance—extinction in these modes becomes what Timothy Morton has termed a "hyperobject," an object so widely distributed in time and space that it cannot be experienced directly but is rather known through data. However, seeing the baiji skeleton or the Chinese paddlefish in the jar or the Carolina parakeet (even a model) brings the hyperobject down to bite size. It puts extinction on an individual scale with a particular name and a particular face. These too are memorials for the dead.

8 *Playful Figures*

"Look closer / Regardez ici," the sign at the Royal Ontario Museum encouraged. Who couldn't help but look? Peering through a small hole on a black divider, the world became full of Carolina parakeets. The nearby information board explains that "the species was extinct by the 1930s," but extinction had turned into abundance through that hole. Through mirror placement, one parakeet had turned into a flock. There were parakeets as far as the eye could see—up, down, left, right. The parakeets went on forever.

Extinction is rightfully talked and written about as serious business. It may be difficult to envision talking about the tragic loss of life-forms in anything but a tragic mode. But as literary scholar Ursula Heise argued in her brilliant book *Imagining Extinction*, perhaps alternatives like comedy need to be considered. Playfulness, absurdity, and ironic frivolity are appropriate

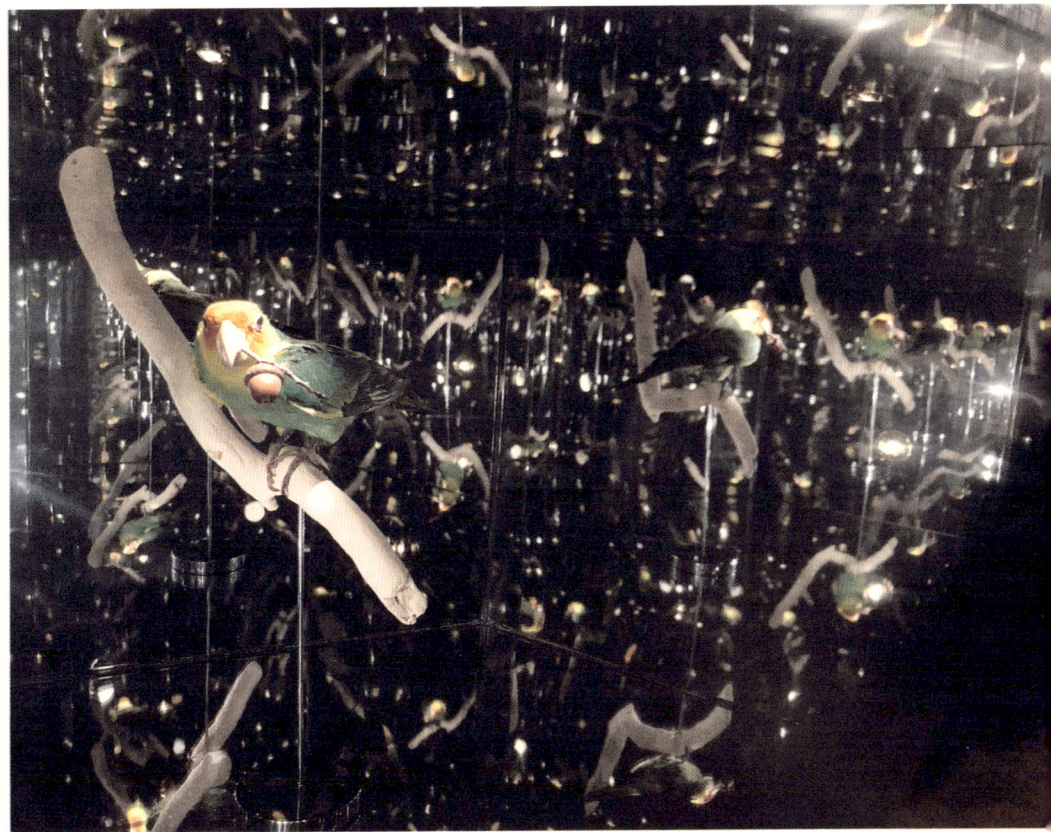

FIG. 64 An imaginary Carolina parakeet flock, Royal Ontario Museum, Toronto, Canada, 2019.

engagement with environmental problems, as shown by Nicole Seymour in *Bad Environmentalism* and Alenda Chang in *Playing Nature*. Play is a valuable part of our modern cultural response to the growing ecological crisis. Play is active, inviting participation in the situation or game and asking for the involvement of the participant's imagination.

Putting an extinct parakeet in a mirrored box to create the illusion of abundance is a playful response to extinction. It invites the visitor in and shows them a world of wonder and joy in imagining a flock. This does not mean that it is not a sad story when the visitor realizes that Carolina parakeets are not in flocks anymore, but it lets them come to that realization in a playful way. We should not think of play as just something children do; instead, play is a natural and cultural category of human behavior at all ages.

In the mid-twentieth century, many natural history museums shifted from galleries filled only with stuffed animals to interactive exhibitions. The educational mission of the museum was reframed with new pedagogy concepts that stressed explorative, active learning in line with newer science center models. Exploratory displays involve visitors, particularly children and youth, to increase understanding and interest in the life sciences. They also appeal to curiosity and wonder through activity.

Many extinction exhibitions draw on elements of play and interactivity to engage visitors in the extinction tragedy. These display techniques can be grouped into different forms from most to least complex: augmented reality, multimedia sound and video, simple interactives, and artistic interpretation. How might these engage the visitor's sense of play to communicate about extinction?

My teenage daughter and I were standing in front of the Steller's sea cow skull in the Endangered and Extinct Species Gallery in the Muséum national d'Histoire naturelle in Paris. The huge skull belongs to the first mammal to become extinct in North America in the modern era. The Steller's sea cow (*Hydrodamalis gigas*), named for the biologist Georg Wilhelm Steller, who encountered it in 1741 on the Bering Strait, was a huge marine mammal measuring up to thirty feet in length and weighing more than an elephant. They were floating in the shallows munching on seaweed when Steller and the sailors of Captain Vitus Bering's ship *St. Peter* found them. The crew, who were stranded and trying to build a boat to sail back to Russia, ended up killing sea cows to eat. But the sea cow would have survived as a species if the crew had not ended up bringing otter, fox, and seal pelts home with them. These sparked the attention of men looking for money in the fur trade. When fur trappers followed in the wake of Bering's expedition, the sea cow became a handy and easily acquired food source. The last of the species was killed in 1768, only twenty-seven years after Steller's fateful encounter.

As my daughter and I looked at the skull, it suddenly appeared to come out of the case, floating before our eyes. It rotated so we could see the whole thing, even the part normally hidden in the glass case. It then moved behind us and we turned to follow it. The rest of the bones grew out of the skull so that a full skeleton was now in the air. Then the skeleton gained its skin and began to swim the length of the gallery. An adorable baby sea cow soon joined in the swim. They were so huge and

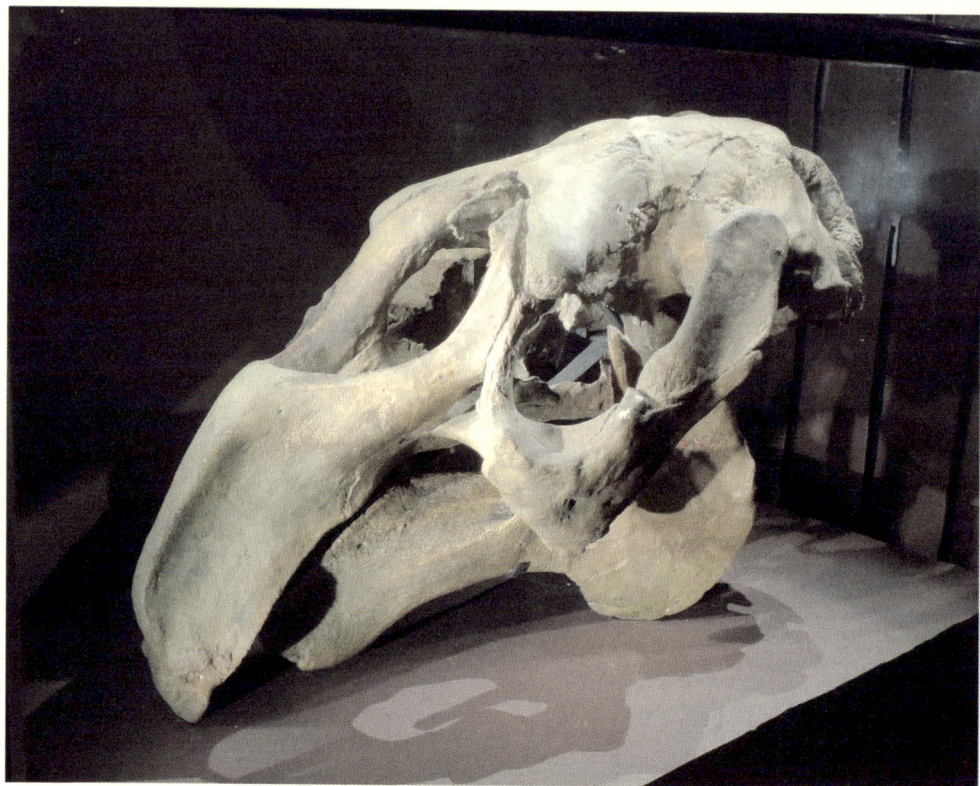

FIG. 65
Steller's sea cow skull,
Muséum national
d'Histoire naturelle,
Paris, France, 2022.
Photograph by Marion
Jørgensen. Used with
permission.

yet graceful, the sound of the waves pulsing as mother and calf frolicked in the seemingly wet air.

We were experiencing *REViVRE*, an augmented reality experience of extinct animals created by the company Saola for the museum. Saola wanted to create an experience in which "visitors will be able to see these fantastic animals moving around before their very eyes for the first time since their extinction." Equipped with an AR smartglasses headset, visitors move through the gallery stopping at different marked extinct species remains. The glasses are such that the visitor can see the room and the physical remains, but in addition, the digital appears on top of the physical objects at various points in the tour. The digital animals are animated in different ways: the thylacine walks on top of the case where the mounted thylacine stands, a barking quagga herd occupies the room facing the stuffed quagga, the elephant bird chick hatches out of the egg on display. In total, there were eight modern extinct species brought to life, one Pleistocene extinction, and a dung beetle that was thought to be extinct but then rediscovered. As the visitor watches these animals at life size, the narrator, who the visitor hears via the headset, explains the history of each species and why it became extinct. The extinction message is couched in emotional terms at the end of the AR experience as an "extraordinary and sad" journey through time and space. The visitor sticks out their hand, a passenger pigeon alights on it, and the narrator asks them to do differently than humans of the past.

All of the animals were animated in 3D in their natural size and moved as if alive, but they were still in the museum room. My teenage daughter remarked that "You weren't in their habitat; they were in yours." It mattered that the AR experience was done in the halls of the museum. It could not have been done the same way elsewhere, because then the visitor wouldn't have had the actual animal remains to interact with.

The amazing Steller's sea cow portrayal in *REViVRE* depended on the move from bones to skin: the skull before your eyes in the museum case became a skeleton and then a fleshy body. This reanimation of the dead skull into a swimming sea cow engaged the viewer, both as a wonder and as a loss. The addition of the calf reminded the viewer of the families that once

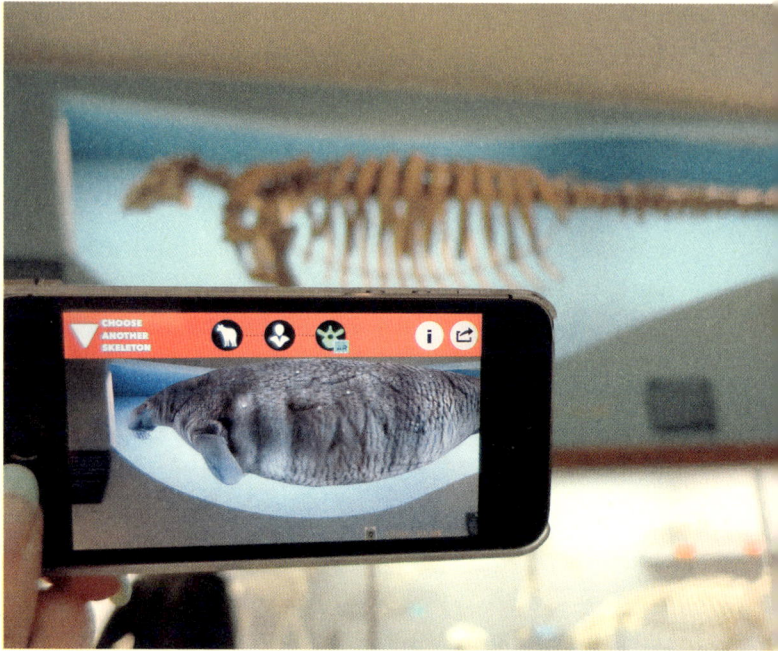

FIG. 66
Steller's sea cow in
the gallery and the
Skin & Bones app,
Smithsonian's National
Museum of Natural
History, Washington,
DC, USA, 2015.

were, an element that is often missed in natural history museums. That sea cow mother and baby seemed alive for those moments as they playfully swam through the hall.

Similar to *REViVRE*'s location-specific interface to turn museum objects into living animals, the Smithsonian's National Museum of Natural History created an AR phone app called *Skin & Bones* back in 2015. The application featured thirteen species, ten augmented reality experiences, information about the animals, videos of scientists who study the species, and four interactive activities. The "skeleton works" option would animate the skeleton that the viewer was standing in front of in the Bones Hall exhibition. The Bones Hall opened in the 1960s as a comparative anatomy exhibit focusing on structures like legs or wings. Some of the AR animations kept the skeleton form and focused on body physics like how movement works. But the Steller's sea cow, which was the only extinct animal in the app, got its flesh back similar to the sea cow in *REViVRE*.

If you held your phone camera with the app at the Steller's sea cow, magically the skeleton that was seen through the lens now had skin and wrinkles, flippers and whiskers. It was a completely different animal than the skeleton. Now the body didn't just disappear after the rib cage and the neck wasn't skinny. It had volume and mass that was missing before. It looked like it might swim away.

In their *Skin & Bones* visitor survey in 2015, Diana Marques and Robert Costello found that visitors were impressed with the recontextualization of the bones. One visitor remarked, "I think it's easy to look at a bunch of bones and forget about the animal that's underneath, or on top I guess . . . visuals to make it more real as opposed to skeletons which disassociate you from nature, from reality." The Steller's sea cow was identified as a favorite by some visitors, one remarking that "I had no idea what it would have looked [like] alive until I had the app—it was extinct but we actually got to see it." According to these visitors, the AR allowed a vision of the specimen that was not possible with the bones alone, which were not able to convey the animal's appearance. The "aliveness" of AR is its primary draw.

Making the extinct come to life can even be done on a much smaller scale than either *REViVRE* or *Skin & Bones*. The Mauritius Natural History Museum installed *Dodo Expedition AR* in the museum in 2021. They set up an open habitat diorama with an older dodo model on a small stage with leaves on the ground and a background showing a forest. By downloading

FIG. 67
The *Dodo Expedition
AR* app in use,
Mauritius Natural
History Museum, Port
Louis, Mauritius, 2022.

the *Dodo Expedition AR* app on your phone, you could point it at the diorama and the single dodo suddenly became a family of birds. These AR birds reflect the latest ideas about the dodo's body configuration and coloring. The birds squeak and squawk as they walk through the rustling leaves. They move and interact with each other. According to the press release on the Mauritius Museums Council website, the "app is all about harnessing the power of Augmented Reality to transport users into the lost world of the extinct Dodo." With AR, the digital animal can be animated to behave as it would have in life in a re-created environment.

These AR experiences, whether on an individual's phone or a boutique headset, are playing with extinction. They make the extinct animal come alive through twenty-first-century technology. Animals can move and act in a way that feels to the watcher as if the live animal is in the room. These applications are, however, expensive to develop. Luckily there are other simpler (and cheaper) multimedia modes of bringing the extinct to life.

UNANSWERED CALLS

A tape reel in a box with a pink label doesn't seem like an exciting object at first, but picking up the speaker under it, the visitor to the British Library's special exhibition *Animals: Art, Science, Sound* was in for a haunting experience. Playing through the speaker was the song of the last Kauaʻi ʻōʻō (*Moho braccatus*).

FIG. 68
The original tape with the recording of the call of the Kauaʻi ʻōʻō, *Animals: Art, Science, Sound* exhibition, British Library, London, UK, 2023.

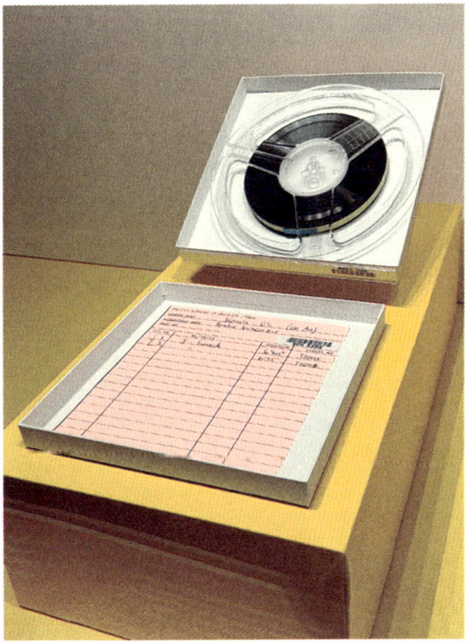

The Kauaʻi ʻōʻō was one of several honeycreepers on the Hawaiian islands when the Polynesians arrived there between 940 and 1130 AD. Each island in the chain had endemic bird species—the same evolutionary radiation that Charles Darwin had observed in the Galapagos. The Kauaʻi ʻōʻō was found only on its eponymous island. It was a small bird, just over eight inches long, with black feathers on head, wings, and tail and a small patch of yellow feathers near the legs. Like all the other ʻōʻō species, deforestation, the introduction of mammalian predators, and new mosquito-transmitted diseases caused its decline. By the 1960s, the number of Kauaʻi ʻōʻō was down in the thirties. By 1981, there was one pair left. The last female bird disappeared during a hurricane, leaving the lonesome male whose solitary song was recorded by John Sincock in 1983. It was a mournful song. Mostly a six-note pattern, four rising notes and two falling. "Where did you go, my love?" you can hear him sing. When this individual bird died, the entire *Mohoidae* family—all of the ʻōʻō—were extinct. In October 2023, the US Fish and Wildlife Service finally took the Kauaʻi ʻōʻō off the endangered species list, admitting that it has been lost completely.

The Indigenous Hawaiians had a close relationship to ʻōʻō. They caught the birds to pluck their yellow feathers, which were woven into the ʻahu ʻula (a cloak or cape) worn by nobility. All of the birds that contributed the yellow feathers became extinct in the twentieth century. Gitte Westergaard has argued that ʻahu ʻula break down the barriers between cultural and natural history museums. She considers these as biocultural artifacts that must be interpreted within the extinction history of the bird as well as the near extinction of the culture that created them. In the American Museum of Natural History, for example, one Hawaiian feathered cape in the ethnographic gallery has a label noting that the mamo (*Drepanis pacifica*), which was used for the yellow feathers, is extinct. As the feathers went, so went the songs.

Sounds like the ʻōʻō call are powerful multimedia instruments of play. The visitor literally presses "play" to hear it. And when hearing it, you cannot help but think about the tragedy of a song being sung when there is no one left to hear it.

Another use of sound that strikes a similar dissonant chord can be heard in the Royal Belgian Institute of Natural Science in Brussels. In the *250 Years of Natural Sciences* exhibition, under the banner labeled "1871," there is a plain dark wooden room to enter. Nearing the entrance of the room, a sound comes seemingly out of nowhere. It is the bleating of sheep. Footsteps in the grass. Then a gunshot and a high-pitched howl. Quickening of the footsteps toward a whimpering animal. Another shot. Silence.

This is the soundtrack for the thylacine room. Its thirty seconds plays whenever a person comes near the entrance. A taxidermy preparation of this "victim of prejudice," as the exhibit was titled, is in a glass case at one end of the room, around a corner. The visitor hears the sound before seeing the animal. But turning the corner, it is clear what was being shot and killed. This particular thylacine, however, was not shot—it was tame and given to the museum in 1871 as a live gift from Tasmanian naturalist Morton Allport. Although it wasn't the one being shot, it stands in for other thylacines that were. The text bemoans the "imbecilic civilization" with "irrational prejudice" that wiped out the thylacine. The display contains many more photographs than other thylacine exhibits (except on Tasmania itself): there is a thylacine strung up, a carcass on a hunter's lap, a mother thylacine with three pups, and the famous endling with gaping mouth.

One of the informational boards titled "posthumous homage" notes that among the few thylacine remains are "some sad photographs and even, the final irony, a number of films of this fine animal in a cage of chicken wire." Outside the room on the right, a video screen loops the film of the last thylacine in a cage. If someone has just entered the room, you can hear the soundscape of the hunt as the background to this silent film of the thylacine pacing.

The museum made an amazing choice to create a soundtrack

for this encounter. It made it impossible not to think of the thylacine as being hunted down and shot. Its whimpers wrench the heart. The curators were playing with media to make the thylacine's death even more tragic than it appears on paper. A fitting, playful, posthumous homage.

PUSH HERE

Many natural history museums are old; their collections were started in the 1800s. But in the twenty-first century, these museums have often been retooled to embrace modern pedagogy techniques of interaction to spur on visitor curiosity. Sometimes these are with digital screens, but often they are simple buttons and levers.

At the Kelvingrove Art Gallery and Museum in Glasgow, a digital screen sits next to the great auk: "*The Tale of Scotland's Last Great Auk*. Press to Start." This interactive component appears as a book with pages that the reader flips by pressing the corner of each page to turn to the next. Each two-page spread has some text (you can read or press the button to hear it out loud) on the left page and a cartoon illustration on the right. It begins, "Once upon a time not very long ago, lived a bird called the great auk. You can see one standing here in front of you. This is the very true and very sad story of the last great auk in Scotland." It tells the history of three men who rowed out to an uninhabited rocky island of St. Kilda to catch seabirds for food. The men end up stuck on the island in a storm and take shelter in a small stone hut. After three days of the storm, the men decide that the bird, which has been crying out the whole time, must be a witch. They take it outside and kill it. The story ends, "From that day to this, no-one has ever seen a great auk

FIG. 69
Great auk with the
digital storybook
about the last Scottish
great auk, Kelvingrove
Art Gallery and
Museum in Glasgow,
UK, 2022.

in Scotland. There is no happy ending to this story. Today there are no great auks living anywhere in the world. Humans have wiped out the great auk."

The digital book includes interactive games, like matching the cartoon drawing to the correct bird image (the correct image for the great auk is the museum's stuffed specimen), pressing items to get additional information (one of these shows the various reasons great auks were hunted), and pressing the auk in the picture to hear it cry. *The Tale of Scotland's Last Great Auk* and its interactive games invite children to play with extinction. It is a real-life ghost story—one that is both entertaining and

FIG. 70
The author's daughter
Lina interacting with
river otters, Ehime
Prefectural Science
Museum, Niihama,
Japan, 2023.

scary. The children get to interact with the story that makes
funny sounds. Yet they are not given a watered-down version
of the story. The ironic frivolity of their play is matched by
the absurdity of the story of the last great auk, which hunters
thought was a witch.

Play engages visitors, particularly young ones, with extinct natural history specimens. For example, the Japanese river otter exhibition in Ehime had two interactive elements. One case had a button to push that changed the lighting so the specimen's coat appeared darker and more brown. The text explained that when alive, the otter had dark-brown fur, but this one's fur changed after death to a white color, so the light would correct it. The other was a piece of otter fur that you could touch. The label explained that this was Canadian otter fur rather than Japanese river otter fur, which is understandable since those are extinct and the fur is too precious, but it was still meaningful to be able to have physical contact with fur. These exhibits embrace physically pushing or touching to focus attention on the animal.

I had not expected play when I visited the Cape lion at the Clifton Park Museum in Rotherham, UK, one of few Cape lions on public exhibit. Known as Nelson, the lion is the museum's most famous artifact. A sign on the wall explains that Nelson is an extinct species of lion with a darker mane than other types and gives some basic information about lion behavior. Then it recounts the specific history of Nelson as a specimen: he was part of a German traveling menagerie until he was sold in 1859 to Carl Hagenbeck, who then sold him on to the Zoological Society of London in 1862 for their London Zoo. After Nelson died, a local nobleman bought him. In 1946, Sir John Whitaker of nearby Retford gave the lion, as well as the rest of his natural history collection, to the museum on long-term loan. In 1970, Whitaker's heir made the donation permanent.

The amazing thing about Nelson, who also serves as the museum's mascot, is that he sits in the Africa-themed children's playroom. The room has a large tree that kids can tuck themselves into and a zebra rocking horse to ride. Children's colored pages of Nelson as cartoon mascot hang on the wall. Some quite frankly racist depictions of native Africans in primitive attire sit above the coloring table. And there is Nelson in the back in a wooden case underneath a picture of the Cape's Table Mountain. On the right of Nelson's case, there is a panel of buttons. One of these has a picture of a lion, and if you press it, a lion loudly roars. The large sign on the wall at child's height explains, "When lions **roar**, the noise they make is so loud that it can be heard 8 kilometers away. If Nelson roared in this room you would be able to hear him in Meadowhall."

Nelson is a museum favorite, and he is everywhere. The find-an-object activity sheets passed out for children are called "Explore with Little Nelson," who according to the sheet is "the cousin of Big Nelson the African Cape Lion." Coloring pages with a cartoon Nelson are available. Nelson's Cub Club invites parents with young kids aged one to four to explore the museum and participate in fun activities.

Nelson was located in a hallway near the stairs until 2003, when he was moved for building refurbishment. When he came back, he entered "The Lion's Den," as the room is referred to. In a newspaper article for the sixty-fifth anniversary of the lion coming to the museum, senior officer of the museum Rachel Reynolds remarked, "Everybody loves Nelson and whenever people visit the museum, the first thing they usually do is visit the Lion's Den." Nelson is a playful artifact that visitors identify childhood with and make memories with. Although the Cape lion is extinct, Nelson is very much alive.

Nelson the Cape lion in a playful environment offers an opportunity for children to engage with animals, even extinct ones, as part of their understanding of environments far from

FIG. 71
Nelson the Cape
lion in the playroom,
Clifton Park Museum,
Rotherham, UK, 2022.

their own daily life. This kind of child's play may sound frivolous and flippant, and former Keeper of Palaeontology at the Natural History Museum (London) Norman MacLeod has argued that gaming trivializes extinction. I disagree. In his "Pedagogy of Play," Roberto Farné argued that "it is in play that the most authentic dimension of the pedagogy of activism is expressed. To play means to be in the game and to act, passivity is incompatible with play." He advocated embracing spontaneous and natural play as learning moments for "political education" for children and adults alike because they have to be active and creative. Recent research in teaching and learning for environmental citizenship has demonstrated that learning activities can indeed contribute to raising student awareness of environmental issues as well as their ability to make decisions that better the environment, which is one of the most critical "political education" components of our age.

WEAVING LOST THREADS

Artistic interventions within natural history museums are also a mode of playing with extinction. Museum studies scholar Sarah Wade has argued that displaying handicrafts like knitting and crochet in natural history museums has the potential to engage visitors with species and habitat loss. Craftivism, the combination of craft and activism to address social and political issues, is a playful way to engage visitors in environmental problems. Examples of the craftivist endeavors she discusses are crochet coral reefs to mark habitat loss, Stitch-A- Squid to engage visitors in learning about the deep sea, and artist Ruth Marshall's knitted pelts to raise awareness of poaching and the wildlife trade.

Marshall's *Tasmanian Tiger #3* (2015) was displayed in the Grant Museum of Zoology as part of the *Strange Creatures: The Art of Unknown Animals* exhibition. Marshall had knitted a thylacine pelt and stretched it across a frame made of string and sticks. The knitted artwork was based on specimen AMNH35504 at the American Museum of Natural History, identified as "female, New York Zoological Society, 26 November 1912." The pelt was displayed alongside other thylacine artifacts—an articulated skeleton and a biological thylacine pelt preserved in liquid. The interpretative text for the case, written by Wade, noted the craftivist potential of Marshall's "knitting to build awareness about hunting and extinction" while at the same time "elicit admiration at the skill involved to create them, perhaps evoking wonder in viewers." Playing with extinction through craft—using the form but not the material of the extinct—unsettles the typical museum extinction display. It asks the viewer to imagine the crafted work as an organic body, splayed out on a frame.

Brandon Ballengée's *RIP Passenger Pigeon, after John James Audubon* (2021), part of his Frameworks series (2006–), similarly plays with extinction. This work is at the Musée de la Chasse et de la Nature in Paris in a room on the theme of the disappearance of wild species. Ballengée, an artist and biologist, had taken a reproduction of Audubon's iconic passenger pigeon illustration from *Birds of America* and carefully excised the birds. The cutouts create an absence. The branches on which the birds had perched are still there, but are now empty.

Extinction as a void, the creation of a hole in the fabric of nature, is made manifest in Ballengée's work. In the artist statement on his website, Ballengée places Frameworks as a commentary on the ongoing biodiversity crisis and sixth great extinction: "Responding to this cataclysm, I physically cut images of missing animals from historic prints and publications printed at the time in history when the depicted species became extinct." This results in a "Framework of Absence," according to the artist. Many of Ballengée's works use original prints he has acquired over the years, such as his *RIP Pied or Labrador Duck: After John James Audubon* (1856/2007), which used an original 1856 Royal Octavo printing. After the image is cut out, Ballengée burns it and gathers the cremation remains. The black urn on top of the fireplace in the Musée de la Chasse et de la Nature contains the ashes of the cutout passenger pigeon image.

Cutting out an image, particularly when it is a valuable early printing, unsettles the artwork. In their discussion of Ballengée's series, Valérie Bienvenue and Nicholas Chare wondered if reactions to the works were "generated by the loss of the animals or the loss of the image." If the response is because of the artwork as a work of art, perhaps it is problematic. After all, as Gordon Sayre notes, artworks like Audubon's were generated by and for the capitalist system that caused the destruction of the very birds Audubon drew. The critique is sidestepped with *RIP Passenger Pigeon, after John James Audubon* (2021), since it actually isn't an old drawing: it is a reproduction of an old drawing. Yet it still manages to provoke the viewer to consider the cost of loss and what is absent. It leaves a haunting blank in the frame.

The addition of modern artworks like these in natural history museum settings offers a different way of knowing extinction than the scientific discourse can. Marshall and Ballengée turn the artifact on display into something unexpected, provoking an emotional reaction. In both cases, the absence of the extinct—by being replaced by string or by being cut out of the picture—is turned into a presence. The extinct ghost is here in the room even if we hadn't noticed it before.

FAIR PLAY

Encountering playful ghosts is entertaining. They move or squawk or roar. They have soft fur or fur of threads. They suddenly have bodies or are unexpectedly absent. It is perhaps the surprise of it all, not knowing what would happen, that is attractive. I think about the time when a model of the ivory-billed woodpecker at the Houston Museum of Natural Science started unexpectedly hammering on a tree—I jumped back and then I laughed. Several minutes later when it did it again and I saw another startled visitor who laughed, I knew I wasn't alone. Just like me, that visitor went up to the board and read about what it was. The excitement drove the curiosity to know.

To end the chapter, I'll add a word of caution that playing with extinction doesn't always work. "The thylacine's last stand" at Smithsonian's Museum of Natural History asks visitors to "PUSH to light up the extinct thylacine" next to a red button with an exclamation point. The display portrays the thylacine's extinction as a battle between the thylacine and the dingo, rather than between thylacine and humans. The dingo is seen in ambient light, but when the button is pushed, the thylacine, which is behind an opaque screen, is lit up. This makes it visible, while the dingo simultaneously is not. It is a somewhat bizarre choice that the Smithsonian made to narrate the extinction of the thylacine in mainland Australia thousands of years ago, rather than on Tasmania, where they existed at the time of European colonization. This does a disservice to the thylacine's story and downplays the incredibly destructive role of White settlers on the fauna across Australia.

The Smithsonian thylacine was a female that had been captured in the wild in 1902. It turned out that it was pregnant and gave birth en route to the United States. The female lived for two years at the zoo; two of the cubs survived to adulthood. After death, the female was prepared by a taxidermist and is the one on display today at the museum. This story is completely

Australia
An Extinction Story
What happened when these two predators faced off?

Australian Eucryptus
blandiana

How the Dingo
Beat Out the Thylacine

When the first wild dingoes arrived with Asian sailors...
the thylacine was...
predator. The...
size and often...
prey. Eventual...
because of its com...

The thylacine's last stand

absent in the display. Not all play with extinction is equally valuable. Play that downplays too much context can subtract rather than add.

Yet this encounter with an extinct ghost at play is not the typical one. More often than not, the playful encounter serves to give the ghosts voices and movements that they are otherwise missing. This is not done as a way of denying their extinction, but, rather, as a demonstration of what has been lost by letting the extinct be present, and playful.

FIG. 72 (*facing*)
An Extinction Story in which the thylacine appears as a shadow behind the dingo, Smithsonian's National Museum of Natural History, Washington, DC, USA, 2019.

9 *Invisible Dead*

I had come to Harvard to see the first North American mammal encountered by the crew of Christopher Columbus in the waters of the West Indies and the second marine mammal to go extinct in the modern era: the Caribbean monk seal (*Monachus tropicalis*). I had been looking for remnants of the animal from the time before it became extinct. I had found scientific drawings, historical photographs of taxidermy mounted animals, a couple photographs of live animals in an aquarium in 1910, and even a set of photographs from a scientific specimen collection hunt in 1900 capturing the brutal slaying of dozens of individuals. There are hundreds of specimens in collections worldwide, with the biggest numbers at the Smithsonian in Washington, DC, and the American Museum of Natural History in New York. But seeing a Caribbean monk seal on display had eluded me.

A male Caribbean seal (*Monachus tropicalis*) obtained at the Triangles by Henry L. Ward in 1886.—The blotches are not the result of any skin disease but are patches of old hair that has not yet been shed

The public could have encountered the Caribbean monk seal at museums in the past. The holotype specimen (the one used to describe the species; it had been collected in the 1840s) was seen by visitors to the British Museum's natural history division up until the 1880s. Another one was on display at Mexico's Museo de Historia Natural de Tacubaya in the 1880s, but it was moved in 1915 to the collections of the Museo de Historia Natural del Chopo and was lost in a fire in 1979. A 1936 guide to the Harvard Natural History Museum noted that the "almost, if not quite, extinct" Caribbean monk seal on display in the South America Room was "very rare in collections." I found a photograph on the website Dinopedia, a Fandom site spotlighting extinct animals, taken by a Harvard museum visitor relatively recently (mid-2010s) of the animal on display

FIG. 73
Photograph of a mounted male Caribbean monk seal specimen in the American Museum of Natural History collection. Published in Frederic A. Lucas, "The Seal Collection," *Natural History* 24, no. 5 (1924): 592.

and then located another photograph on a different site. The mounted Caribbean monk seal (labeled as the West Indian monk seal) was on the light-green floor of the case under the New World monkeys. Here was my best chance to see one of these extinct animals in person.

But when I arrived at the museum, I discovered that the Caribbean monk seal had been taken to storage around 2017, when the display cases were updated. A river otter and a harp seal were located where the Caribbean monk seal once stood. The extinct Caribbean monk seal had experienced a double extinction—it had disappeared from public view.

The list of extinctions since 1500 is quite long—776 animal species had the IUCN "extinct" classification as of 2023 (297 mollusks, 159 birds, 85 mammals, 82 insects, 81 bony fishes, 36 amphibians, 32 reptiles, 4 worms)—and very few remains of those species are available for the public to see anywhere in the world. These missing bodies, and their stories, cannot be easily chalked up to rarity. After all, the Smithsonian has pages of entries in its specimen ledger for its Caribbean monk seal specimens, including taxidermy mounts, study skins, and skeletal remains, and uses them for cutting-edge genetics research, yet no Caribbean monk seal is on display. There is of course the issue of space, but how a museum uses space is a choice: Why should a common harp seal be on display but not a Caribbean monk seal? There are also questions of preservation, but even a poor specimen might be worth displaying to tell its story like the pathetic Carolina parakeet in Leuven.

What extinctions are *not* encountered by the museum visitor? In this chapter, I want to think with and on behalf of the extinct that are typically invisible in the museum. This includes reptiles and insects, as well as species that have very few extant remains.

In 1986, Alain Delcourt, a scientist at the Muséum d'Histoire naturelle de Marseille, discovered the body of an unnamed species of gecko in the museum stores. The giant gecko had never been scientifically described, and no other specimens were known. This gecko was huge: over 50 percent longer snout-to-vent than any other known gecko, with its length at 370 mm compared to the previous record 240 mm. Aaron M. Bauer and Anthony P. Russell, who first described it, named the species *Hoplodactylus delcourti* after its discoverer. The unexpected discovery and work to solve the mystery of the gecko remains has even been captured in a children's book, *La malédiction du gecko*, by Laurence Talairach in her Enquêtes au Museum series.

Unfortunately there was no collecting information about the specimen—no collector name, no place of collection, and no date. This meant that the researchers had to make associations between this gecko and known (much smaller) geckos to assign it to a family. Bauer and Russell identified it morphologically as belonging to the New Zealand geckos of the genus *Hoplodactylus*, which also allowed them to associate this species with a Māori tradition of giant forest lizards called *kawekaweau*. This association with New Zealand and Māori tradition led to the lending of the specimen to the National Museum of New Zealand in 1990 for an exhibition called *Forgotten Fauna: New Zealand's Amphibians and Reptiles*. Its display prompted sighting reports to come in, but the museum scientists who investigated the reports did not find convincing evidence of the presence of *Hoplodactylus delcourti*, so it was presumed extinct.

Although Bauer and Russell had decided that *Hoplodactylus delcourti* was from New Zealand, other evidence pointed to an origin on New Caledonia, a Pacific island group claimed

FIG. 74 Model of Delcourt's gecko, Musée d'Histoire naturelle de Lille, Lille, France, 2022.

as a French territory. Not only had nineteenth-century French natural history collectors been very active on New Caledonian islands, but the region is also home to some of the Carphodactylini group of lizards, the same group that *Hoplodactylus* belongs to. DNA analysis published in 2023 showed that the giant gecko is indeed more related to existing New Caledonian geckos rather than New Zealand geckos. The origin of the gecko would thus appear to be New Caledonia, although it is not closely related enough to be in the same genus as known lizards. The scientists have suggested renaming it *Gigarcanum delcourti* (for its giant size).

The Musée d'Histoire naturelle de Lille has a gorgeous model of the Delcourt's gecko in a display case about extinction. Next to the more famous dodo (also a model) and below the watchful eyes of a pair of taxidermied huia, the gecko suns itself on a rock. The skin wrinkles and folds across the textured body. The glass eyes give the "illusion of life," as a nearby display about the craft of taxidermy states. The museum clearly labels the gecko as a model and notes on the label the species' discovery at its sister French regional natural history museum in Marseille.

The Delcourt's gecko is a museum celebrity in Marseille—it is the basis of the Muséum d'Histoire naturelle de Marseille's logo, and a photo of the preserved body features on the museum's website as a treasure. The museum's extinction display is tucked away in a back corner. An elephant bird skeleton and egg, thylacine body and film of the last thylacine, and composite dodo skeleton are labeled as examples of "animals of yesterday, now gone." The Delcourt's gecko is not there. When I inquired about the whereabouts of the gecko, I was told that the specimen is kept locked away in storage. The Musée d'Histoire naturelle de Lille had a model on display, and I was surprised that the Muséum d'Histoire naturelle de Marseille had nothing.

What is ironic is that the Delcourt's gecko had hidden unnoticed in the drawers of the museum in Marseille for about 150 years, came to light for a moment, and then returned to the darkness. But it left a ghostly imprint in the process. The stylized gecko on the glass door and the ticket is a reminder of the invisible gecko.

While dinosaurs are public favorites to see on a day out at the museum, amphibians, reptiles, and fish that have become extinct in the modern era are rarely displayed for the public. Amphibians, reptiles, and fish are particularly difficult to preserve by regular taxidermy processes because of the skin qualities and tendency to shrink, so models like the Delcourt's gecko in Lille can be used instead. Some specimens appear in wet preparation, like the Chinese paddlefish in London or the giant lizards in Copenhagen. Yet, most of the time, museums do not bother to put these types of animals on display. They are rendered invisible.

ISLAND SPIRITS

Even if you travel to the seventy-plus museums in the appendix, most species can only be encountered in one of them. While a few animals like thylacines, passenger pigeons, and moas are on display in many museums, I saw forty-nine species in the appendix just once in all my travels. That demonstrates the incredible diversity of extinct species and the rarity of encountering a particular animal if you do not visit many places. Many, many of these single sightings came from islands. Ascension Island crake. Cuban macaw. Curio's giant rice rat. Guadalupe storm petrel. New Caledonian rail. 'ō'ō from Oʻahu and Kauaʻi. The list goes on.

Islands are well known both for the radiation and multiplication of species (think of Darwin's finches from the Galapagos) and for the loss of species through extinction. Islands are confined geographies that typically have few predators and species that have evolved into small niches. It is no wonder that local island inhabitants tended to succumb pretty quickly when humans arrived carrying rats, snakes, pigs, machetes, and guns. At times the animals were specifically hunted, but many times, their end was unintentional and their extinction was collateral damage of European settlement.

We often know little about these island disappearances, and few bodies remain. Take St. Helena's giant earwig (*Labidura herculeana*), which I have only encountered as an image on a stamp at the British Library. The giant earwig was a giant for its type, reaching up to eight cm long. The art for the stamp, which was part of a set on the island's invertebrates, was created in 1995, over twenty years since the last earwig had been seen. It was officially declared extinct in 2014. There are a few preserved specimens of this giant: London (2), Copenhagen (1), Harvard (1), and Leiden (1). None are on display.

Or take the Saint Lucia rice rat (*Megalomys luciae*). It sits on its haunches against a black backdrop near its extinct relative the Martinique rice rat at the Muséum national d'Histoire naturelle

FIG. 75
Saint Lucia rice rat, Muséum national d'Histoire naturelle, Paris, France, 2022.

in Paris. This is one of only two known specimens of the Saint Lucia rice rat, and the only one that is stuffed and displayed for the public. The extinction of the giant rice rats has to be understood within the context of island colonization by Europeans. The introduction of black rats in the Caribbean along with the transformation of the landscape into sugarcane plantations spelled doom for the endemic rice rats. Only a handful of them were collected by scientists before they were gone.

Island extinctions are pervasive across the globe and yet nearly invisible, except for the ubiquitous dodo. Even other Mauritian species are overshadowed by the dodo. While the small Mauritian flying fox (*Pteropus subniger*) is on display in Paris near the rice rat, there was no example of this extinct bat that once lived on the Mascarene Islands on display in a Mauritian museum. A commentator writing in 1772 had already noted their declining numbers because of human predation.

FIG. 76
Small Mauritian flying
fox, Muséum national
d'Histoire naturelle,
Paris, France, 2022.

This flying fox species became extinct in the early 1800s. While several museums include them in their catalogs, this haunting nocturnal animal is only on display in one.

INVISIBLE INVERTEBRATES

Invertebrates, including crustaceans, mollusks, and insects, are the most infrequently displayed extinct animals, especially in terms of a percentage of extinct animals. There is a real bias in natural history museums toward animals that are large or cute, and preferably both. Yet insects do get displays and even whole galleries in natural history museums, particularly as rooms designed for interactive children's exhibitions. These do not, however, feature extinct insects.

The Xerces blue butterfly (*Glaucopsyche xerces*) is one of the few extinct insects on display at multiple museums. The butterfly had inhabited the San Francisco area, living in sandy dune habitats. But in 1849, in the wake of the California gold rush, the human population around the Bay exploded. Urban development disturbed and overtook the sand dunes, causing the extinction of the deerwood plant, which was the preferred larval host plant for the Xerces blue. In 1875, naturalist H. H. Behr sent specimens of the butterfly to Herman Strecker, curator of entomology at the California Academy of Sciences, along with a letter stating: "L. Xerces will be in the box. It is now extinct as regards the neighborhood of S. Francisco. The locality where it used to be found is converted into building lots, and between German chickens and Irish hogs no insect can exist besides louse and flea." Collecting rare and now extinct species was seen as a praiseworthy feat in another of Behr's letters: "It is wonderful how many of our species that formerly were quite common are nearly or entirely extinct, for instance, Lycaena Xerces." Behr was wrong about the species being extinct in the 1870s, but it was only a matter of time. The last specimens were collected by entomologist W. Harry Lange at San Francisco's Presidio military base in 1941.

The California Academy of Sciences, which accepted the Xerces blues sent by Behr, has a display case on the butterfly's extinction, but like the last quagga's de-extinction story, the information stresses hope. A related butterfly, the silvery blue, is being studied as a surrogate butterfly for reintroduction to the San Francisco dunes, according to the sign. This replacement butterfly is shown in a small habitat re-creation on a plastic plant, while the Xerces blues are mounted in a frame. The narrative is filled with hope and action: "We can learn from past extinctions . . . Though we can't bring back the extinct Xerces butterfly, we may save its threatened relatives, like the Mission blue, by replanting native flowers they depend on." By including the stories of the silvery blue and Mission blue butterflies along with the Xerces blue, the museum was able to create a sufficiently large display that highlighted these small butterflies.

Size is typically a problem for displaying insects. Although the Xerces was considered a lepidopterist prize, it is dwarfed when put on display with other extinct animals. Such was the case at the Royal Ontario Museum in Toronto, Canada, where the butterfly is tiny compared to the other extinct animals and it doesn't look particularly blue (something noticeable also about the one in Exeter).

FIG. 77 (*facing*) Xerces blue butterflies (from top to bottom: male top, female top, underside), California Academy of Sciences, San Francisco, CA, USA, 2022.

Because of their small size and the unfamiliarity of many visitors with the numerous species, crustaceans, mollusks, and insects do not get the same level of detail as the larger animals when displayed. For example, although there are many species

FIG. 78
A very tiny Xerces
butterfly displayed with
much larger ivory-
billed woodpecker,
passenger pigeon, and
thylacine skull under a
"Gone Forever" label,
Royal Ontario Museum,
Toronto, Canada, 2019.

of Hawaiian snails that have gone extinct, they are typically grouped and displayed without species names. At Kelvingrove Art Gallery and Museum, Glasgow, UK, three shells represent the extinct Hawaiian land snails (*Carelia spp.*). At Manchester Museum, collector's boxes with several snail species have handwritten information about the specific species and collection location, but the case label just says "Hawaiian tree snails." There are actually over 750 recognized species of Hawaiian land snails, and as Thom van Dooren recounts in his book *A World in a Shell*, there are many more waiting to be described from the millions of museum specimens. Almost all of these new species will already be extinct. Invasive species have taken their toll on Hawaiian snail numbers, but so did shell collectors, particularly in the late 1800s and early 1900s. These hobbyists amassed thousands of shells in personal collections, often

FIG. 79
Species of Partula land snails, Museo Nacional de Ciencias Naturales, Madrid, Spain, 2022.

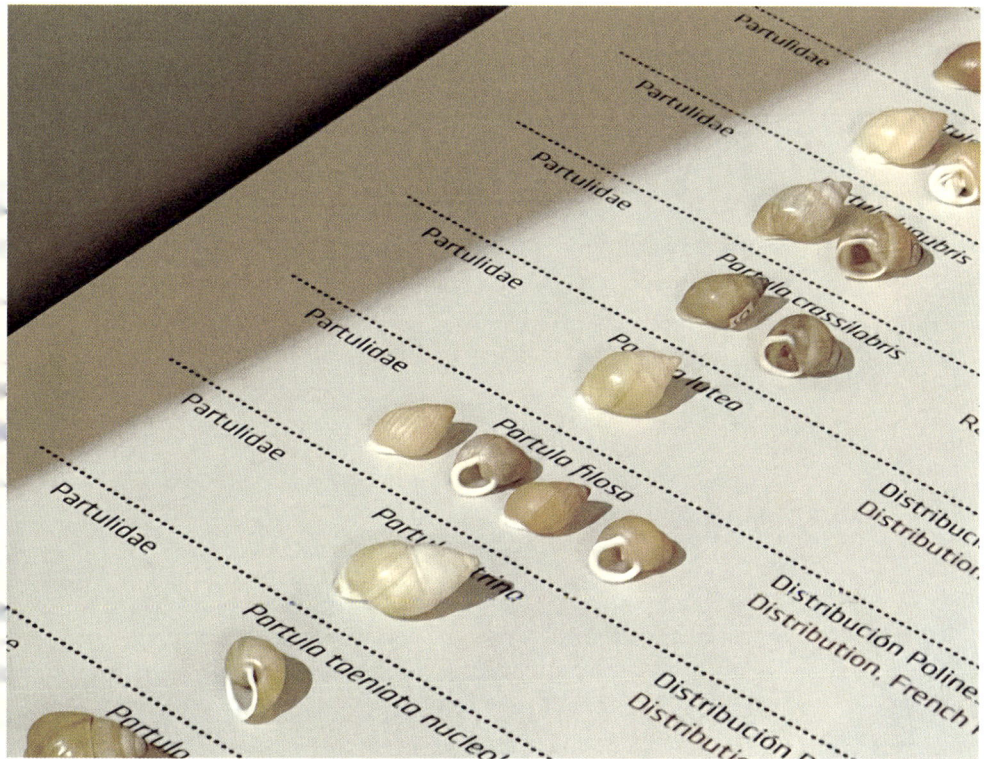

collecting the rarest snails and probably contributing to their eventual extinctions. Some of these animal remains made it into museum malacology collections like the Bishops Museum, which has a staggering six million Pacific island snail shells. As van Dooren observes in discussing the plethora of Hawaiian snail extinctions, extinction is not a singular phenomenon and each of these extinct snail species had a way of living, impacting entangled lives and landscapes. But that individuality and specificity is erased with these invertebrate extinction displays.

Because of the numbers of invertebrates going extinct, they could be vehicles for communicating the enormous toll of extinction. At the Museo Nacional de Ciencias Naturales in Madrid, the extinction section features some standard subjects—thylacine, dodo, giant moa, and great auk—alongside an atypical display of "the extinction of continental mollusks" with shells of nineteen named species of *Partula* land snails. Of the eighty or so different species of *Partula* from the Pacific islands, over half have become extinct in the twentieth century. The extinctions have been caused by exotic species introductions; some, like the giant snail *Achatina fulica*, were intended as pest control but ended up devouring the native land snails. The display probably does not make much of an impression on the visitor, with the shells on the edges of shadows in the case, but it is a good example of displaying the scale of extinction and naming it.

SEALED FATE?

With all the extinctions around us, perhaps it is not surprising that many go unseen and unremarked in museums. Perhaps part of the human problem with extinction is that it is distant and nonpersonal. Extinction becomes a rare event that happened to iconic dodos or thylacines. The hundreds of other commonplace species that may not be as pretty or well known disappear. But is it fated to be this way?

Although Caribbean monk seals cannot be encountered in a museum now, you can come face-to-face with a Japanese sea lion that became extinct at just about the same time. The Japanese sea lion (*Zalophus japonicus,* also known as the Dokdo sea lion) inhabited the coastal areas of the Japanese archipelago and the Korean peninsula. In the 1800s and early 1900s, sea lions were harvested commercially for oil. The hunting pressure was immense: Japanese fishermen commercially hunted 3,200 sea lions in 1904, but by the early 1910s the numbers were down to a couple of hundred because the animal population had plummeted. The last reports of Japanese sea lions in the area were fifty to sixty individuals seen on the islets group of Takeshima/Dokdo between Japan and the Korean peninsula in 1951. A few animals had been captured for use in circuses and zoos up to 1941. The Tennōji zoo in Osaka had live Japanese sea lions in the 1940s. After the last died, they replaced the exhibit with California sea lions.

The small natural history museum at the Tennōji zoo shows off specimens that once lived in the zoo, and one of those is a Japanese sea lion. It was sitting on a low black wooden display platform on the floor. You can tell it was a sea lion rather than a seal by the ears—unlike seals, sea lions have external ear lobes. Its body was covered in smooth, almost golden hair. Near the specimen are a series of information boards that tell a story about the species from right to left. In this narrative, sea lions were hunted for oil but were also feared by hunters. A large male known as Daiou Liangko was particularly dangerous

FIG. 80
Japanese sea lion,
Tennōji Zoo Museum,
Osaka, Japan, 2023.

because he was so territorial. He was eventually killed, stuffed, and put on display at the Sahimel Shimane Nature Museum of Mt. Sanbe, where I saw him later. Two color photos of the huge bull sea lion accompany an admonition: "Don't forget that there was an animal called Japanese sea lion."

The command to not forget the Japanese sea lion seems appropriate in light of its cousin the Caribbean monk seal's story. They were both hunted for oil. They probably became extinct about the same time in the first half of the twentieth century. A few were stuffed and put on display. But the Caribbean monk seal is no longer visible to the public. The Japanese sea lion still

is. Stuffed Japanese sea lions are on display in several museums around the island, including the Sahimel Shimane Nature
Museum of Mt. Sanbe (the display includes the mighty Daiou
Liangko with two others), Shimane University museum (this
specimen has been designated as a Prefectural Natural Monument), Aquas Shimane Aquarium, the National Museum of
Territory and Sovereignty, and Kyūshū University Museum
(this specimen was a captive animal kept at Fukuoka aquarium). Sea lion models are also on display in museums on the
island of Ulleungdo, South Korea. While natural history museums are not the only place in which remembering or forget-
ting animals happens, they are a central node in remembrance
practices for the extinct. It is the place that the bodies of the
dead can be made visible.

Persistent Presence

It would be easy to associate the contemporary extinction crisis with disappearance—after all, animal populations are rapidly disappearing—but I hope this book has revealed the ongoing presence of the extinct. In the halls of the museum, body parts, artistic works, and words re-present the extinct. They are not the same beings that they once were, but still they are present. Visitors have a chance to come face-to-face with these beings.

After visiting so many museums around the world and encountering so many ghosts behind glass, I am left weighing the value of displaying extinction. After all, these extinct animals have been the victims of terrific violence. The animals whose bodies are now preserved with chemicals and put in front of the visitor were almost always intentionally killed by hunters or collectors. Families were ripped apart. Lifeways were ended. What is the value of seeing the remnants of that?

The value lies in understanding and engaging with our on-going environmental crisis, especially for reckoning with the violence of extinction. Scholarship on museums and human tragedy affirms that there is a deep value to such engagement. Sociologist Amy Sodaro has investigated memorial museums exhibiting the atrocities of genocide, terrorism, and human rights abuses. She concluded that museums such as the US Holocaust Memorial Museum and the Museum of Memory and Human Rights in Chile have three main functions: truth-telling about history and preserving the past; serving as a solemn space of remembrance to help heal and repair; and instilling visitors with a "never again" ethic. While belief in progress has been thoroughly shaken in our age, these museums still maintain that "confronting the negative past can lead to a better future, reflecting broader assumptions about the ethical duty to remember."

Sodaro's observations also apply to extinction, arguably the greatest atrocity committed by humans on nonhumans. First, the natural history museum, in particular, preserves the remains of extinct animals as relics of the past. This is the institution best equipped to take care of these remains for long-term preservation. With these remains, museums are equipped to tell truthful stories about the past—stories that do not shy away from the inherent inequity of power in the relations between humans and animal species that have led to the latter's demise. Second, museums serve as spaces of remembrance, whether through lists of the extinct, artistic visions of the lost, or glimpses of landscapes past in dioramas. Whether the visitor stands in front of stilled taxidermied remains, or watches them come to life in AR, museums offer opportunities for contemplation about life and death. Finally, museums are well placed to communicate a "never again" ethic. It is not that extinctions will not happen—I dread to think how many species went extinct over the course of writing this book—but there is an opportunity to ask people to think and act differently.

Museums are incredibly well equipped to allow for haunting by ghosts of the extinct. Natural history museums have the remains, the places for display, and audiences eager to hear the ghost stories. Other museums can also get into the action, sharing art and literature that likewise contains the ghostly presence of the extinct. I encountered many school groups during my visits, and it gives me hope. Children are fascinated by the wonder of nature and strongly sympathize with beings that have been wrongly treated. They are an audience willing to be moved by extinction stories, if given the chance to encounter them. By facilitating encounters with extinct animals, museums of all kinds can instill in visitors a sense of wonder and excitement in seeing animals that they can never witness running or crawling or slithering in the wild.

We need to seek out encounters with the ghosts of extinction in museums—and we need to take time to relish the encounter when it happens. When I visited museums with my family, they were often impatient as I took my time in front of every extinct specimen. But taking time was critical. Slowing time in the museum is one of the more important things that we can do within the museum space in the face of rapid environmental change. I needed to take time to experience the wonder and the pain of extinction. I needed to contemplate meeting a being that only inhabits a museum gallery. Every single one of the encounters was meaningful and magical.

My hope is that readers of this book will be inspired by these encounters with the extinct. I want you to take time with the remains of the extinct next time you visit a natural history museum. I hope that through this book you are better equipped to read displays for what is there and what isn't. Think about the treasures and the play, the relations and memorials, the forms and heritage. Allow the ghosts to come out from behind the glass.

Acknowledgments

This book could not have been possible without the funding necessary to visit museums all over the world. For that I particularly thank the Research Council of Norway, which funded the project "Beyond Dodos and Dinosaurs: Displaying Extinction and Recovery in Museums" (no. 283523) and my portion of the Joint Programming Initiative for Cultural Heritage project "Extinction as Cultural Heritage? Exhibiting Human-Nature Entanglements with Extinct and Threatened Species" (no. 296921), and the Department of Cultural Studies and Languages at the University of Stavanger. Thanks are also due to the many people and organizations that invited me to talk at various events over the last seven years, as I not only got to present my in-progress ideas but likely also took a side trip over to a local museum or two to look for extinction stories.

Warm thanks to Gitte Westergaard, who came with me for fieldwork on Mauritius and pushed my thinking with her research on island extinctions and biocultural artifacts, and Verity Burke, who as the postdoctoral fellow on "Beyond Dodos and Dinosaurs" pushed me to think about copies, digital existence, and authenticity. And finally, a word of thanks to my family, who saw many of these exhibits with me, for letting me see the exhibits through other eyes. I will cherish the memories of my daughter Marion experiencing the AR display in Paris alongside me and my daughter Lina playfully enjoying the Japanese river otter room.

Ghosts Encountered

These are the museums I visited for research on this book and the species that became extinct because of humans that I encountered in each. All of these remains were on public display at the venue. Most often these remains were labeled as extinct, although in some cases current scientific consensus does not support that designation.

Since museum exhibits change over time, the date of my encounter is indicated (sometimes a museum was visited more than once, so it has multiple dates). If the encounter was as part of a special exhibit shown only for a limited engagement, the name of the exhibition is noted. It is possible that there were more extinct animals on display at the museum at the time of my visit, but I am listing here only those that I encountered.

If the specimens of a given species were not solely taxidermy mounts, the type of specimen is indicated as taxidermy mount (T), skeletal (S), unmounted skin (U), model and casts (M), egg (E), wet preparation (W), art/image (A), shell (L), or other (O). The number of individuals is also indicated if more than one of the same type was on display.

AFRICAN, ASIAN, AND AUSTRALASIAN COLLECTIONS

Australia

Australian Museum, Sydney
FEBRUARY 5, 2016
Macquarie Island red-crowned parakeet
Thylacine (T, S, W)

National Museum of Australia, Canberra
FEBRUARY 16, 2016
Thylacine (S)

Tasmanian Museum and Art Gallery, Hobart
FEBRUARY 8, 2016
Macquarie Island red-crowned parakeet (S)
Thylacine (A, T, 4S, U)

China

Shanghai Natural History Museum, Shanghai
MARCH 28, 2018
Baiji (S)
Timeline of extinct species with images of many species

Japan

Aquas Shimane Aquarium, Hamada
JUNE 10, 2024
Japanese sea lion

Ehime Prefectural Science Museum, Niihama
APRIL 6, 2023
Honshū wolf (S)
Japanese river otter (6T, 4M, U, S)

Hall of Inspiration, University Museum of University of Tokyo, Tokyo
APRIL 7, 2023
Moa (E)

Intermediatheque, Tokyo
APRIL 4, 2023
Elephant bird (M of S, M of E)

Special Exhibit *Birds in Paradise*
Carolina parakeet (A)
Moa (M of E)

National Museum of Nature and Science, Tokyo
APRIL 7, 2023
*Elephant bird (E) *in gift shop*
Honshū wolf (S, T)

Japanese river otter (S)
Thylacine

National Museum of Territory and Sovereignty, Tokyo
Japanese sea lion (T, A, O)

Sahimel, the Shimane Nature Museum of Mt. Sanbe, Sanbe
JUNE 10, 2024
Honshū wolf (A)
Japanese crested ibis (extinct in Japan) (U)
Japanese river otter
Japanese sea lion (3T, O)
Japanese serow (extinct in prefecture)
Oriental stock (extinct in Japan) (U)
Snail Margaritifera laevis (A)

Shimane University Museum ASHIKARU, Matsue City
JUNE 11, 2024
Japanese sea lion

Tennōji Zoo Museum, Osaka
APRIL 5, 2023
Japanese sea lion

Mauritius

Black River Gorges National Park Pétrin Visitors' Centre, Le Pétrin
APRIL 8, 2022
Plegma dupontiana (L)
Tropidophora carinata (L)

Blue Penny Museum, Port Louis
APRIL 6, 2022
Dodo (A)

Frederik Hendrik Museum, Grand Port
APRIL 8, 2022
Dodo (M, A)
Mauritius blue pigeon (A)
Red rail (A)
Rodrigues solitaire (A)

National History Museum, Mahebourg
APRIL 8, 2022
  Dodo (S, A)
  Giant land tortoise (Saddle-backed and Domed giant together) (S)
  Red rail (A)
  Rodrigues solitaire (S)

Natural History Museum, Port Louis
APRIL 7, 2022
  Dodo (S—3 complete plus extra bones, M, O, A)
  Elephant bird (E, S)
  Giant land tortoise (saddle-backed and domed giant together) (S)
  Mauritius blue pigeon (T)
  Red rail (S)
  Rodrigues solitaire (S)

Singapore

Lee Kong Chian Natural History Museum, Singapore
JUNE 24, 2016
  Dodo (M, S)

South Korea

Dokdo Exhibition Hall, Seoul
JUNE 14, 2024
  Dokdo (Japanese) sea lion (A)

Dokdo Museum, Seoul
MARCH 2, 2024
  Dokdo (Japanese) sea lion (A, O)

Dokdo Museum, Ulleung
SEPTEMBER 5, 2024
  Dokdo (Japanese) sea lion (A, O)

Japanese-Style House in Dodong-ri, Ulleung
SEPTEMBER 5, 2024
  Dokdo (Japanese) sea lion (A, O)

National Science Museum, Daejeon
FEBRUARY 28, 2024
Acipenser sinensis (extinct in the country)
Ciconia boyciana (extinct in the country)

Ulleungdo Marine Protected Area Visitor Center, Ulleung
SEPTEMBER 4, 2024
Dokdo (Japanese) sea lion (M, A, O)

Taiwan

National Taiwan Museum, Taipei
OCTOBER 11, 2023
Formosan clouded leopard (2T)

EUROPEAN COLLECTIONS

Austria

Naturhistorisches Museum Wien, Vienna
MAY 24, 2018, AND DECEMBER 28, 2018
Carolina parakeet
Dodo (S, multiple M)
Elephant bird (E, S)
Giant moa (E, S)
Ivory-billed woodpecker (in the endangered section)
Passenger pigeon
Steller's sea cow (S)
Thylacine

Belgium

Gents Universiteitsmuseum, Ghent
OCTOBER 7, 2022
Thylacine

Katholieke Universiteit Leuven Museum voor Dierkunde, Leuven
JANUARY 25, 2023
  Carolina parakeet
  Giant moa (M)
  Passenger pigeon

Royal Belgian Institute of Natural Science, Brussels
OCTOBER 8, 2022
  Thylacine

Denmark

Statens Naturhistoriske Museum, Copenhagen
DECEMBER 14, 2021
  Dodo (S)
  Giant anoles (5W)
  Great auk (2T, E, 4W)
  Thylacine (2S)

Finland

Luonnontieteellinen museo, Helsinki
MAY 3, 2022
  Giant moa (A)
  Javan tiger (S)
  Steller's sea cow (S)

France

Musée de la Chasse and de la Nature, Paris
FEBRUARY 12, 2022
  Passenger pigeon (A)

Musée d'Histoire naturelle de Lille, Lille
OCTOBER 9, 2022
  Delcourt's gecko (M)
  Dodo (M)
  Huia (2T)

Muséum d'Histoire naturelle, Nantes

SEPTEMBER 25, 2024
 Carelia cumingiana snail (L)
 Elephant bird (E, S)
 Gibbus lyonetianus snail (L)
 Great auk
 Passenger pigeon (2T)

Muséum d'Histoire naturelle de Marseille, Marseille

AUGUST 6, 2023
 Dodo (S)
 Eastern moa (S)
 Elephant bird (E, S)
 Moa (E)
 Parantica aspasia butterfly (O)
 Thylacine

Muséum d'Histoire naturelle de Nîmes, Nîmes

JULY 23, 2023
 Aurochs

Muséum national d'Histoire naturelle, Paris

JUNE 29, 2015, AND FEBRUARY 12, 2022
 Barbary lion
 Black emu (S)
 Bluebuck
 Broad-faced potoroo
 Cape lion
 Carolina parakeet
 Corsican deer
 Crescent nail-tail wallaby
 Dodo (M of S)
 Eastern hare-wallaby
 Elephant bird (E)
 Great auk
 Grey's wallaby
 Hawai'i ʻōʻō
 Insects: 8 butterfly species, 1 damselfly, 1 beetle
 Ivory-billed woodpecker

Martinique giant rice rat
New Caledonian rail
Oʻahu ʻōʻō
Passenger pigeon
Pink-headed duck
Quagga
Saint Lucian giant rice rat
Schomburgk's deer
Small Mauritian flying fox
South Island piopio
Southern pig-footed bandicoot
Steller's sea cow (3S)
Thylacine

Special Exhibition REViVRE
Dodo (O)
Elephant bird (O)
Great auk (O)
Passenger pigeon (O)
Quagga (O)
Rodrigues tortoise (O)
Steller's sea cow (O)
Thylacine (O)

Germany

Botanisches Museum, Berlin
NOVEMBER 17, 2014
**Special Exhibit Caucasus. Plant Biodiversity Between
the Black Sea and the Caspian Sea**
Caspian tiger (S)

Museum der Natur Hamburg, Hamburg
SEPTEMBER 23, 2022
Eastern hare-wallaby (O)
List of extinct species from IUCN
Quagga (O)
Thylacine (O)

Museum für Naturkunde Berlin, Berlin
MAY 8, 2014, AND SEPTEMBER 22, 2022
  Dodo (M)
  Huia (2T)
  Quagga
  Spix's Macaw (2T)
  Thylacine

Museum Koenig, Bonn
OCTOBER 19, 2022
  Wolf (marked as locally extinct)

Naturmuseum Senckenberg, Frankfurt
MARCH 6, 2022
  Carolina parakeet
  Dodo (S, M)
  Giant moa (S)
  Great auk (T, S)
  Heavy-footed moa (S)
  Huia (2)
  Passenger pigeon
  Quagga
  Thylacine

Italy

Museo Civico di Storia Naturale di Milano, Milan
OCTOBER 24, 2019
  Elephant bird (E)
  Giant moa (S)
  Great auk

The Netherlands

Naturalis, Leiden
OCTOBER 10, 2021, AND SEPTEMBER 1, 2022
  Dodo (M)

Special Exhibit *Nature's Treasure Trove—200 Years of Naturalis*
  Cape lion
  Dodo (S)
  Elephant bird (E)
  Great auk (T, E)
  Quagga

Natuurhistorisch Museum, Maastricht
JANUARY 25, 2023
  Passenger pigeon (2)

Teylers Museum, Haarlem
OCTOBER 10, 2021
  Dodo (S)
  Elephant bird (M)

Special Exhibit *Vogelpracht*
  Carolina parakeet (T, A)
  Great auk (A)
  Heath hen (A)
  Ivory-billed woodpecker (A)
  Passenger pigeon (A)

Norway

Museum Stavanger, Stavanger
AUGUST 12, 2019
  Javan tiger

Naturhistorisk museum, Oslo
NOVEMBER 12, 2014, AND JULY 9, 2019
  Elephant bird (E)
  Great auk
  Thylacine

Universitetsmuseet i Bergen, Bergen
JULY 13, 2021, AND MAY 25, 2022
  Carolina parakeet
  Dodo (M)
  Giant moa (S)

Passenger pigeon
Thylacine

Poland

Muzeum Przyrodnicze, Wrocław
OCTOBER 7, 2021
    Carolina parakeet
    Elephant bird (E)
    Great auk
    Huia
    Javan lapwing
    Passenger pigeon
    Thylacine

Spain

Museo Nacional de Ciencias Naturales, Madrid
MARCH 5, 2022
    Dodo (M)
    Freshwater bivalves (L collection of 3 species)
    Giant moa (E)
    Great auk
    Island land snails (L collection of 19 species)
    Thylacine

Sweden

Evolutionsmuseet, Uppsala
OCTOBER 15, 2014
    Great auk

Naturhistoriska riksmuseet, Stockholm
MAY 7, 2019, AND NOVEMBER 2, 2021
    Bluebuck
    Cuban macaw
    Elephant bird (E)
    Giant moa (S)
    Great auk

Hawaiian snails (L)
Hoopoe starling
Huia (2T)
Ivory-billed woodpecker
Quagga
Steller's sea cow (S)
Thylacine

United Kingdom

Bristol Museum and Art Gallery, Bristol

MAY 9, 2017, AND NOVEMBER 10, 2018
    Burchell's zebra subspecies
    Dodo (M)
    Thylacine

SEPTEMBER 25, 2019
Special Exhibit *Extinction Voices*
    Dodo (M)
    Galapagos giant tortoise
    Thylacine

British Library, London

JULY 3, 2023
Special Exhibit *Animals: Art, Science, Sound*
    Carolina parakeet (A)
    Falkland Islands wolf (A)
    Ivory-billed woodpecker (O)
    Kaua'i ōō (O)
    St. Helena giant earwig (A)
    Thylacine (O)

Clifton Park Museum, Rotherham

MAY 6, 2022
    Cape lion

Grant Museum of Zoology, London

JUNE 12, 2015, AND MAY 4, 2017
    Dodo (S)
    Pyrenean ibex (S)

Quagga (S)
Thylacine (T, S, W, O)

Horniman Museum and Gardens, London
DECEMBER 12, 2015
   Dodo (M)
   Passenger pigeon (2)

Hunterian Museum, Glasgow
JUNE 9, 2022
   Bluebuck (S)
   Dodo (M of head)
   Elephant bird (S, E)
   Giant moa (S)
   Jamaican giant galliwasp (W)
   Thylacine

Kelvingrove Art Gallery and Museum, Glasgow
JUNE 9, 2022
   Dodo (small scale M)
   Great auk
   Hawaiian land snails (3L)
   Huia
   Madeiran large white butterfly
   Passenger pigeon

Leeds City Museum, Leeds
SEPTEMBER 14, 2016
   Dodo (S)
   Heavy-footed moa (S)
   Thylacine (S, O)

Manchester Museum, Manchester
NOVEMBER 23, 2018
   Carolina parakeet (U)
   Dodo (M)
   Great auk (S)
   Hawaiian tree snails (L)
   Huia (2)
   Ivory-billed woodpecker (U)

Moa (S)
Paradise parrot (U)
Passenger pigeon (U)
South Island piopio (U)
Steller's sea cow (S)
Thylacine (S)

National Museum of Scotland, Edinburgh
JANUARY 21, 2020; MAY 5, 2022; AND MARCH 1, 2023

Ascension Island crake (S)
Aurochs (S)
Baiji (M)
Bluebuck (O)
Bush wren
Curio's giant rice rat
Dodo (M, S)
Golden toad
Guadalupe storm petrel
Huia (2T)
Mauritius blue pigeon
New Zealand quail
Passenger pigeon
Pink-headed duck
Polynesian tree snail, Partula turgida
Quagga
Rodrigues solitaire (S)
Schomburgk's deer (O)
Steller's sea cow (S, M)
Thylacine

Special Exhibit *Audubon's Birds of America*
Carolina parakeet (A, T)
Ivory-billed woodpecker (A)
Passenger pigeon (A, T)

Natural History Museum, London
OCTOBER 30, 2013; DECEMBER 12, 2015; AND SEPTEMBER 28, 2022

Carolina parakeet
Dodo (S, A, M)

Great auk (2T)
Huia
Thylacine

Special Exhibit *Our Broken Planet*
    23 species of bees extinct in UK
    Chinese paddlefish
    Thylacine

Natural History Museum at Tring, Tring
MAY 6, 2017
    Burchell's zebra subspecies
    Carolina parakeet
    Dodo (M)
    Elephant bird (S)
    Giant moa (S)
    Great auk
    Passenger pigeon
    Quagga
    Thylacine (T, S)

Special Exhibit *Dodos: Old Bird, New Tricks*
    Dodo (S, M, A)

Oxford University Museum of Natural History, Oxford
MAY 6, 2017, AND OCTOBER 12, 2018
    Dodo (S, M)
    Elephant bird (S, E)

Royal Albert Memorial Museum and Art Gallery, Exeter
NOVEMBER 4, 2018
    Carolina parakeet
    Eskimo curlew (U)
    Hawaiian tree snails (L)
    Heath hen (U)
    Huia (2T)
    Ivory-billed woodpecker (U)
    Passenger pigeon
    Xerces blue butterfly

University Museum of Zoology, Cambridge
MAY 7, 2022
    Dodo (S, 4M)
    Elephant bird (E)
    Great auk (S)
    Hawaiian honeycreepers (2T, 4U)
    Huia (2T)
    Moa (S)
    Rodrigues solitaire (S)
    Steller's sea cow (S)
    Thylacine (S)

Wollaton Hall, Nottingham
MAY 6, 2022
    Dodo (2M)
    Moa (S)
    Paradise parrot
    Passenger pigeon

NORTH AMERICAN COLLECTIONS

Canada

Archives of Ontario, Toronto
NOVEMBER 7, 2019
Special Exhibit *Animalia, Animals in the Archives*
    Passenger pigeon (3U, O)

Canadian Museum of Nature, Ottawa
NOVEMBER 9, 2019
    Great auk (S)
    Passenger pigeon

Royal Ontario Museum, Toronto
NOVEMBER 12, 2019
    Blue pike
    Carolina parakeet
    Deepwater cisco
    Dodo (S)

Freshwater mollusk Epioblasma arcaeformis
Freshwater mollusk Pyrgulopsis nevadensis
Giant moa (S, E)
Golden toad
Ivory-billed woodpecker (2)
Labrador duck
Passenger pigeon
Thylacine (S)
Xerces blue butterfly

United States of America

Academy of Natural Sciences of Drexel University, Philadelphia, Pennsylvania

OCTOBER 4, 2023

13 species of birds from the Mascarene Islands, only dodo named (A)
Eskimo curlew (2)
Giant moa (A)
Great auk
Labrador duck (2)
Passenger pigeon (group)

American Museum of Natural History, New York, New York

NOVEMBER 14, 2015, AND NOVEMBER 14, 2021

Dodo (M of S)
Labrador duck (4T)
Passenger pigeon (group plus single)

California Academy of Sciences, San Francisco, California

DECEMBER 29, 2022

24-rayed sunstar
California grizzly bear (S)
Ivory-billed woodpecker (2M)
Passenger pigeon (A)
Xerces blue butterfly

Cincinnati Zoo and Botanical Garden, Cincinnati, Ohio

JUNE 12, 2017, AND APRIL 12, 2019

Dusky sparrow
Passenger pigeon (T, M)

Denver Museum of Nature and Science, Denver, Colorado
APRIL 7, 2024
  Carolina parakeet (4T)
  Elephant bird (E)
  Ivory-billed woodpecker (3T)
  Passenger pigeon (15T)

Field Museum, Chicago, Illinois
OCTOBER 28, 2014
  Passenger pigeon (groupT)

Harvard Museum of Natural History, Cambridge, Massachusetts
NOVEMBER 12, 2021, AND MARCH 26, 2023
  24-rayed sunstar (not labeled extinct)
  Elephant bird (E)
  Ivory-billed woodpecker (2T)
  Javan tree shrew (labeled extinct)
  Moa (S)
  Thylacine (T, S)

Houston Museum of Natural Science, Houston, Texas
MARCH 21, 2023
  Carolina parakeet
  Ivory-billed woodpecker (M, T)
  Passenger pigeon

Illinois State Museum, Springfield, Illinois
APRIL 2, 2024
  Carolina parakeet (T, M)
  Passenger pigeon

National Mississippi River Museum and Aquarium, Dubuque, Iowa
MARCH 26, 2017
  Passenger pigeon (2T)

Smithsonian National Museum of Natural History, Washington, DC
MARCH 22, 2015, AND JUNE 6, 2019
  Ivory-billed woodpecker
  Steller's sea cow (S)
  Thylacine

Special Exhibit *Once There Were Billions: Vanished Birds of North America*
    Carolina parakeet (2T, A)
    Great auk (T, E, A)
    Heath hen (2T, A)
    Labrador duck (A)
    Passenger pigeon (3T, A)

Witte Museum, San Antonio, Texas
SEPTEMBER 29, 2019
Special Exhibit *Mythic Creatures*
    Elephant bird (M)

Notes to Sources

PROLOGUE

My thinking about extinction ghosts is inspired by philosopher Vinciane Despret, who works on relations between humans and nonhuman animals but took a side step with her book *Our Grateful Dead: Stories of Those Left Behind*, trans. Stephen Muecke (Minneapolis: University of Minnesota Press, 2021). Working through her own family's history, Despret argues that the dead take on alternative other existences that we encounter through "their potential to act, or rather, to give rise to action, through their capacity to affect us from the 'outside'" (p. 7). This is more than just remembering the dead; it is letting them still live and act in the world. While Despret was not writing about extinction, or even animals, her book encourages us to think about the prolonging of existence for the extinct.

Some scholars have also used the concept of *hauntology* to talk about social or cultural elements of the past in the present, but often their ghost is more metaphorical than physical. French philosopher Jacques Derrida coined the term in *Spectres of Marx: The State of the Debt, the Work of Mourning and the New International*, trans. Peggy Kamuf (London: Routledge, 1994) to talk about the way that Marxism continued to "haunt" modern society. There are many good studies of "hauntology" and the ideas put forward in Jacques Derrida, such as Avery F. Gordon, *Ghostly Matters: Haunting and the Sociological Imagination* (Minneapolis: University of Minnesota Press, 2008). I find particularly useful the approach of Colin Sterling, "Becoming Hauntologists: A New Model for Critical-Creative Heritage Practice," *Heritage & Society* 14, no. 1 (2021): 67–86, in applying the concept to the museum contexts I work in.

For ghost species and the spectral nature of extinction, see Shane McCorristine and William M. Adams, "Ghost Species: Spectral Geographies of Biodiversity Conservation," *Cultural Geographies* 27, no. 1 (2020): 101–115; Adam Searle, "Hunting Ghosts: On Spectacles of Spectrality and the Trophy Animal," *Cultural Geographies* 28, no. 3 (2021): 513–530; and Adam Searle, "Spectral Ecologies: De/extinction in the Pyrenees," *Transactions of the Institute of British Geographers* 47 (2022): 167–183. Art historians Valérie Bienvenue and Nicholas Chare used the word *afterimage* to talk about representations of extinction as a phantom presence in the present in *Animals, Plants and Afterimages: The Art and Science of Representing Extinction*, ed. Valérie Bienvenue and Nicholas Chare (New York and Oxford: Berghahn Books, 2022). I don't adopt the word *afterimage* here because it implies something much more static and passive (something observed or seen) than Despret was talking about. I have been particularly inspired by observations about the proliferation of extinct species in Audra Mitchell, "Beyond Biodiversity and Species: Problematizing Extinction," *Theory, Culture & Society* 33, no. 5 (2016): 23–42.

CHAPTER 1

There is a huge amount of scientific research showing the rapid pace and great extent of animal extinction in the modern age, that is, since 1500. I have drawn the "defaunation of the Anthropocene" from Rodolfo Dirzo, Hillary S. Young, Mauro Galetti,

Gerardo Ceballos, Nick J. B. Isaac, and Ben Collen, "Defaunation in the Anthropocene," *Science* 345, no. 6195 (2014): 401–406. Extinction, however, is not a new concern, as demonstrated by Frederic A. Lucas, "Animals Recently Extinct or Threatened with Extermination, as Represented in the Collections of the U.S. National Museum," *Annual Report of the Board of Regents of the Smithsonian Institution . . . for Year Ending June 30, 1889* (Washington, DC: GPO, 1891). The quote is from page 609.

I have been particularly influenced in my approach to extinction stories by work on extinction situated in environmental history: Mark V. Barrow Jr., *Nature's Ghosts: Confronting Extinction from the Age of Jefferson to the Age of Ecology* (Chicago: University of Chicago Press, 2009); Ryan Tucker Jones, *Empire of Extinction: Russians and the North Pacific's Strange Beasts of the Sea, 1741–1867* (Oxford: Oxford University Press, 2014); Miles Powell, *Vanishing America: Species Extinction, Racial Peril, and the Origins of Conservation* (Cambridge, MA: Harvard University Press, 2016); Peter S. Alagona, *After the Grizzly: Endangered Species and the Politics of California* (Berkeley: University of California Press, 2020); and David Sepkowski, *Catastrophic Thinking: Extinction and the Value of Diversity from Darwin to the Anthropocene* (Chicago: University of Chicago Press, 2020).

There is a growing body of literature on extinction from the perspective of the wider environmental humanities, which I also draw from extensively for this book. Ursula Heise's work is foundational, specifically *Imagining Extinction: The Cultural Meanings of Endangered Species* (Chicago and London: University of Chicago Press, 2016) and her seminal article "Lost Dogs, Last Birds, and Listed Species: Cultures of Extinction," *Configurations* 18 (2010): 49–72. "Extinction Studies" was coined by Deborah Bird Rose, Thom van Dooren, and Matthew Chrulew in their eponymous volume *Extinction Studies: Stories of Time, Death, and Generations*, ed. Deborah Bird Rose, Thom van Dooren, and Matthew Chrulew (New York: Columbia University Press, 2017). For extinction as violence and a withdrawal from social contracts, see K. J. Hernández, June M. Rubis, Noah Theriault, Zoe Todd, Audra Mitchell, Bawaka Country, Laklak Burarrwanga, Ritjilili Ganambarr, Merrkiyawuy Ganambarr-Stubbs, Banbapuy Ganambarr, Djawundil Maymuru, Sandie Suchet-Pearson, Kate Lloyd, and Sarah Wright, "The Creatures Collective: Manifestings," *Environment & Planning E: Nature and Space* 4, no. 3 (2021): 838–863; Noah Theriault and Audra Mitchell, "Extinction," in *Anthropocene Unseen: A Lexicon*, ed. Cymene Howe and Anand Pandian (Goleta: Punctum Books, 2020), 176–182; and Noah Theriault, Timothy Leduc, Audra Mitchell, June Mary Rubis, and Norma Jacobs Gaehowako, "Living Protocols: Remaking Worlds in the Face of Extinction," *Social & Cultural Geography* 21, no. 7 (2020): 893–908.

I also build on "lively ethnographies" as an approach to multispecies stories from Thom van Dooren and Deborah B. Rose, "Lively Ethnography: Storying Animist Worlds," *Environmental Humanities* 8, no. 1 (2016): 77–94; and the lively taxidermy in Alice Would, "Taxidermy Time: Fleshing Out the Animals of British Taxidermy in the Long Nineteenth Century, 1820–1914" (PhD thesis, University of Bristol, 2021). I draw on ideas of the agency of matter from scholarship such as Jane Bennett, *Vibrant Matter: A Political Ecology of Things* (Durham, NC: Duke University Press, 2010); and Timothy LeCain, *The Matter of History: How Things Create the Past* (Cambridge: Cambridge University Press, 2017).

For the history of the natural history museum, see Paula Findlen, *Possessing Nature: Museums, Collecting, and Scientific Culture in Early Modern Italy* (Oakland: University of California Press, 1996); Oliver Cummings Farringdon, "The Rise of Natural History Museums," *Science*, New Series 42, no. 1076 (1915): 197–208 (the included quote is on page 208); and Barbara T. Gates, "Introduction: Why Victorian Natural History?," *Victorian Literature and Culture* 35 (2007): 539–549. On dinosaurs, see Lukas Rieppel, *Assembling the Dinosaur: Fossil Hunters, Tycoons, and the Making of a Spectacle* (Cambridge, MA: Harvard University Press, 2019). For modern animal displays in the natural history museum, I rely on Karen A. Rader and Victoria E. M. Cain, *Life on Display: Revolutionizing U.S. Museums of Science and Natural History in the Twentieth Century* (Chicago: University of Chicago Press, 2014); Liv Emma Thorsen, Karen A. Rader, and Adam Dodd, eds., *Animals on Display: The Creaturely in Museums, Zoos, and Natural History* (University Park: Pennsylvania State University Press, 2013); and Liv Emma Thorsen, "Animal Matter in Museums: Exemplifying Materiality," in Hilda Kean and Philip Howell, eds., *The Routledge Companion to Animal-Human History* (Abingdon: Routledge, 2018), 171–195. Another good collection of essays on taxidermied individual animals and the museum is Samuel J. M. M. Alberti, ed., *The Afterlives of Animals: A Museum Menagerie* (Charlottesville: University of Virginia Press, 2011). The Alberti quote comes from Samuel J. M. M. Alberti, "Introduction: The Dead Ark," in that collection, page 8.

On William Hornaday and the use of taxidermy as a conservation form for natural history museums, see Hanna Rose Shell, "Skin Deep: Taxidermy, Embodiment, and Extinction," in *The Past, Present and Future of Natural History: Proceedings of the California Academy of Sciences*, ed. Alan Levinton (San Francisco: California Academy of Sciences, 2004), 88–112 (the quote is on page 95); and Gregory J. Dehler, *The Most Defiant Devil: William Temple Hornaday and His Controversial Crusade to Save American Wildlife* (Charlottesville: University of Virginia Press, 2013). I have also read his own words on the subject: William T. Hornaday, *Taxidermy and Zoological Collecting* (New York: Charles Scribner's Sons, 1894); and William T. Hornaday, *Our Vanishing Wildlife: Its Extermination and Preservation* (New York: Charles Scribner's Sons, 1913). His quote is from page 1 of *Taxidermy and Zoological Collecting*.

Most critically, this chapter and the book as a whole build on new scholarship on museum displays of extinction. The most influential pieces are Anna Guasco, "'As Dead as a Dodo': Extinction Narratives and Multispecies Justice in the Museum," *Environment and Planning E: Nature and Space* 4, no. 3 (2021) 1055–1076; Dominic O'Key, "Why Look at Taxidermy Animals? Exhibiting, Curating and Mourning the

Sixth Mass Extinction Event," *International Journal of Heritage Studies* 27, no. 6 (2020): 635–653; and all of the articles in the *Museum and Society* journal's special issue "Exhibiting Extinction" from 2022, which I edited. In this chapter I was thinking about two of the special-issue articles explicitly: Miranda Cichy, "Paradise Lost: Encounters with Australia's Extinct Parrot," *Museum and Society* 20, no. 1 (2022): 89–103; and Isla Gladstone and Persephone Pearl, "Extinction Voices, Extinction Silences: Reflecting on a Decolonial Role for Natural History Exhibits in Promoting Thinking about Global Ecological Crisis, Using a Case Study from Bristol Museums," *Museum and Society* 20, no. 1 (2022): 50–70. Michael Blencowe, *Gone: A Search for What Remains of the World's Extinct Creatures* (London: Leaping Hare Press, 2021), is also a compelling read on museum specimens (mostly behind-the-scenes).

My concept of animal remains is taken from Sarah Bezan and Robert McKay, "Animal Remains: An Introduction," in *Animal Remains*, ed. Sarah Bezan and Robert McKay (London: Routledge, 2021), 1–11. The quote is from page 5.

For *affect* as used by ecocriticism scholars, some key works are Alexa Weik von Mossner, ed., *Moving Environments: Affect, Emotion, Ecology, and Film* (Waterloo, Ontario: Wilfrid Laurier University Press, 2014); Alexa Weik von Mossner, *Affective Ecologies: Empathy, Emotion, and Environmental Narrative* (Columbus: Ohio State University Press, 2017); and Kyle Bladow and Jennifer Ladino, eds., *Affective Ecocriticism: Emotion, Embodiment, Environment* (Lincoln: University of Nebraska Press, 2018).

The key works on taxidermy are Rachel Poliquin, *The Breathless Zoo: Taxidermy and the Cultures of Longing* (University Park: Pennsylvania State University Press, 2012); Rachel Poliquin, "Matter and Meaning in Museum Taxidermy," *Museum and Society* 6, no. 2 (2008): 123–134 (quotes from pages 127 and 129); Jane Desmond, "Vivacious Remains: An Afterword on Taxidermy's Forms, Fictions, Facticity, and Futures," *Configurations* 27, no. 2 (2019): 257–266; Mary Anne Andrei, *Nature's Mirror: How Taxidermists Shaped America's Natural History Museums and Saved Endangered Species* (Chicago: University of Chicago Press, 2020); and Alice Would's PhD thesis. For the specific story of the Shanghai Museum, see Li-Chuan Tai, "The Shanghai Museum and the Introduction of Taxidermy and Habitat Diorama into China, 1874–1952," *Archives of Natural History* 48, no. 1 (2021): 111–130.

All the species discussed briefly in this chapter have interesting extinction histories in their own right. See Cichy, "Paradise Lost," for the paradise parrot. For the Japanese wolf's history, see Brett L. Walker, *The Lost Wolves of Japan* (Seattle: University of Washington Press, 2005). Some new scholarship has argued that Indigenous Māori oral tradition remembers the extinction of the moa: Priscilla M. Wehi, Murray P. Cox, Tom Roa, and Hēmi Whaanga, "Human Perceptions of Megafaunal Extinction Events Revealed by Linguistic Analysis of Indigenous Oral Traditions," *Human Ecology* 46 (2018): 461–470. I note that to my knowledge, these findings have not yet been incorporated into museum exhibits.

To understand how some of the choices in how to group specimens create "portraits of extinction," see Dolly Jørgensen, "Portraits of Extinction: Encountering Extinction Narratives in Natural History Museums," in *Traces of the Animal Past: Methodological Challenges in Animal History*, ed. Jennifer Bonnell and Sean Kheraj (Calgary: Calgary University Press, 2022), 371–387.

An early plea for recognizing the educational role of natural history museums is Barton Warren Evermann, "Modern Natural History Museums and Their Relation to Public Education," *Scientific Monthly* 6, no. 1 (1918): 5–36. Dana Fragomeni, "The Evolution of Exhibit Labels," *Faculty of Information Quarterly* 2, no. 2 (2010), provides a nice review of historical labeling and analysis of updated practices. On the value of labels for visitor interaction, see Paulette McManus, "Oh, Yes, They Do: How Museum Visitors Read Labels and Interact with Exhibit Texts," *Curator: The Museum Journal* 32, no. 3 (1989): 174–89.

For the issue of glass cases as mediators, I am drawing on Brita Brenna, "Nature and Texts in Glass Cases: The Vitrine as a Tool for Textualizing Nature," *Nordic Journal of Science and Technology Studies* 2, no. 1 (2014): 46–51; M. Henning, *Museums, Media and Cultural Theory* (Maidenhead: Open University Press, 2006); and Steph Berns, "Considering the Glass Case: Material Encounters between Museums, Visitors and Religious Objects," *Journal of Material Culture* 21, no. 2 (2016): 153–168. The quote is from Berns, page 161.

Contemporary visitor photography practices have been analyzed in Debra Leighton, "In the Frame: Investigating the Use of Mobile Phone Photography," *International Journal of Nonprofit and Voluntary Sector Marketing* 12 (2007): 308–319; and Theopisti Stylianou-Lambert, "Photographing in the Art Museum: Visitor Attitudes and Motivations," *Visitor Studies* 20, no. 2 (2017): 114–137. The Gothenburg study was documented in a conference paper by Thomas Hillman, Alexandra Weilenmann, and Beata Jungselius, "Creating Live Experiences with Real and Stuffed Animals: The Use of Mobile Technologies in Museums."

CHAPTER 2

There is significant literature on the Tasmanian tiger's history, including Robert Paddle, *The Last Tasmanian Tiger: The History and Extinction of the Thylacine* (Cambridge: Cambridge University Press, 2000); and David Maynard and Tammy Gordon, *Tasmanian Tiger: Precious Little Remains* (Launceston: Queen Victoria Museum and Art Gallery, 2014). For historical takes on the animal's pending extinction in the past, I refer to Richard Lydekker, *A Hand-book to the Marsupialia and Monotremata* (London: E. Lloyd, 1896) (the quote is on page 152); and John West, *The History of Tasmania*, vol. 1 (Tasmania: Henry Dowling, 1852) (the quote is on page 323).

For the history of museum collection and display of thylacines, I have relied on Kathryn Medlock, "Exhibiting Extinction: Thylacines in Museum Display," in *Animals, Plants and Afterimages: The Art and Science of Representing Extinction*, ed. Valérie Bienvenue and Nicholas Chare (New York and Oxford: Berghahn Books, 2022), 392–406.

I previously published an article discussing the hunt for thylacines and the uncertainty of extinction, which makes a broader argument about the relationship between absence and presence: Dolly Jørgensen, "Presence of Absence, Absence of Presence, and Extinction Narratives," in *Nature, Temporality and Environmental Management: Scandinavian and Australian Perspectives on Peoples and Landscapes*, ed. Lesley Head, Katarina Saltzman, Gunhild Setten, and Marie Stenseke (Abingdon: Routledge, 2017), 45–58. The article contains references to the materials used to reconstruct the history. The most important primary source is M. S. R. Sharland, "In Search of the Thylacine: Society's Interest in the Preservation of a Unique Marsupial," in *Proceedings of the Royal Zoological Society of New South Wales for the Year 1938–9* (Sydney: Royal Zoological Society, 1939), 20–38. The quote is from page 20. For the debunking of the 2021 citing, see Naaman Zhou, "Wildlife Expert Pours Cold Water on Claims Tasmanian Tiger Family Spotted," *The Guardian*, February 23, 2021, https://www.theguardian.com/australia-news/2021/feb/23/wildlife-expert-pours-cold-water-on-claims-tasmanian-tiger-family-spotted.

The questionable story of the thylacine in 1962 is told in Michael Sharland, *Tasmanian Wild Life* (Melbourne: Melbourne University Press, 1963), quotes on pages 8–9. The recanting of the tale is told in Eric Guiler and Philippe Godard, *Tasmanian Tiger: A Lesson to be Learnt* (Perth: Abrolhos Publishing, 1998). The curator's reflections on the extinction exhibition were published as Florence Raulin-Cerceau, "La salle des espèces disparues: La valorisation des collections," *La letter de l'OCIM* 33 (1994): 54–57, quote from page 57.

For the history of the ivory-billed woodpecker, I have referred to Mikko Saikku, "'Home in the Big Forest': Decline of the Ivory-billed Woodpecker and Its Habitat in the United States," in *Encountering the Past in Nature: Essays in Environmental History*, ed. Timo Myllyntaus and Mikko Saikku (Athens: Ohio University Press, 1999), 94–140; Matthew Gandy, "An Arkansas Parable for the Anthropocene," *Annals of the American Association of Geographers* 112, no. 2 (2022): 368–386 (quote from page 373); Noel F. R. Snyder, David E. Brown, and Kevin B. Clark, *The Travails of Two Woodpeckers: Ivory-Bills and Imperials* (Albuquerque: University of New Mexico Press, 2009); and Ursula Heise, "Lost Dogs, Last Birds, and Listed Species: Cultures of Extinction," *Configurations* 18 (2010): 49–72. The claim of a sighting in 1962 is included in Tim Gallagher, *The Grail Bird: The Rediscovery of the Ivory-Billed Woodpecker* (Boston and New York: Houghton Mifflin, 2006). The recording in 1971 was published in John William Hardy, "A Tape Recording of a Possible Ivory-Billed Woodpecker Call," *American Birds* 29, no. 3 (1975): 647–651. The claim of a new sighting in 2004 was published in John W. Fitzpatrick et al., "Ivory-Billed Woodpecker (*Campephilus principalis*) Persists in Continental North America," *Science* 308 (2005): 1460–1462. The 2004 claim is covered in Erik Stokstad, "Gambling on a Ghost Bird," *Science* 317, no. 5840 (2007): 888–892. The ivory-billed woodpecker delisting comments are hundreds of pages long.

The quotes in this chapter from McCorristine and Adams come from pages 106 and 109 of "Ghost Species."

CHAPTER 3

The natural history museum as temple of nature is discussed in J. B. Bullen, "Alfred Waterhouse's Romanesque 'Temple of Nature': The Natural History Museum, London," *Architectural History* 49 (2006): 257–285. The quotes are from pages 257 and 271. For more on the architecture of these museums, see John Holmes, "Science and the Language of Natural History Museum Architecture: Problems of Interpretation," *Museum & Society* 17, no. 3 (2019): 342–361. The Natural History Museum (London) has issued a brochure for its traveling exhibition *Treasures of the Natural World*, which I used for the description of the Barbary lion in the gallery. Naturalis issued a catalog for the 200-year exhibition: Eulàlia Gassó and Tiny Monquil-Broersen, eds., *Van onschatbare waarde: 200 jaar Naturalis* (Zutphen: Walburg Pers, 2020).

Deidre Jackson, *Lion* (London: Reaktion Books, 2010), gives an overview treatment of lions as entangled with humans. For early descriptions of the Cape lion, I have relied on Charles Hamilton Smith, *Introduction to Mammalia*, Naturalist's Library, vol. 15, ed. William Jardine (London: Chatto & Windus, 1842); and Edward Griffith, *General and Particular Descriptions of the Vertebrated Animals, Arranged Conformably to the Modern Discoveries and Improvements in Zoology*, vol. 1, Order Carnivora (London: Baldwin, Cradock, and Joy, 1821). The quote from Griffith is on page 30. For a biography of Charles Hamilton Smith, see *Dictionary of National Biography, 1885–1900*, vol. 53, Smith–Stanger (New York: Macmillan, 1898), 24. The foremost authority on the Cape lion, Vratislav Mazak, who published mainly in the 1960s, considered that "we have nowadays no more doubts that the extinct Cape lion represented a valid geographical subspecies and its systematic status grew clear": Vratislav Mazak, "Preliminary List of the Specimens of *Panthea leo melanochaitus* Ch. H. Smith, 1842, Preserved in the Museums of the Whole World in 1963," *Zeitschrift für Säugetierkunde: im Auftrage der Deutschen Gesellschaft für Säugetierkunde e.V.* 29, no. 1 (1964): 52–58. For the lions and menageries, see Mia Uys and Sandra Swart, "Big Cat Acts and Big Men: Performing Power and Gender in South Africa's Circus Industry, c. 1888–1916," *Early Popular Visual Culture* 18, no. 3 (2020): 283–304; and Caroline Grigson, *Menagerie: The History of Exotic Animals in England, 1100–1837* (Oxford: Oxford University Press, 2016). For a discussion of the dynamics of African imperial hunting and environmental change, see William Beinart, "Empire, Hunting and Ecological Change in Southern and Central Africa," *Past & Present* 128

(1990): 162–186. For the collection of Cape lions for museums, see Mazak, "Preliminary List of the Specimens," which provides descriptions and photographs of each specimen. Laura M. F. Bertens and Ann Marie Wilson, "Wonder, Empire, Science: The Quagga and Other Extinctions on Display at Naturalis," *Museum and Society* 20, no. 1 (2022): 33–49, discussed the Cape lion at Naturalis. Quote is from page 42.

The list of extinct South African Cape animals comes from Q. B. Hendley, "The Late Cenozoic Carnivora of the South-Western Cape Province," *Annals of the South African Museum* 63 (January 1974): 1–369, 20. The artwork by Antoine-Louis Barye is in the Morgan Library and Museum, call number B3 FS 029 01, accession number 1996.66.

I previously wrote about encountering bluebucks in Dolly Jørgensen, "Portraits of Extinction: Encountering Bluebuck Narratives in the Natural History Museum," in *Traces of the Animal Past: Methodological Challenges in Animal History*, ed. Jennifer Bonnell and Sean Kheraj (Calgary: University of Calgary Press, 2022), 371–387. A general description of bluebucks is in Philip Lutley Sclater and Oldfield Thomas, *The Book of Antelopes*, vol. 4 (London: R. H. Porter, 1899–1900). The bluebuck and identification of "true" bluebuck specimens have received recent scholarly attention: Elisabeth Hempel, Faysal Bibi, J. Tyler Faith, James S. Brink, Daniela C. Kalthoff, Pepijn Kamminga, Johanna L. A. Paijmans, Michael V. Westbury, Michael Hofreiter, and Frank Zachos, "Identifying the True Number of Specimens of the Extinct Blue Antelope (*Hippotragus leucophaeus*)," *Scientific Reports* 11 (2021): article 2100, https://doi .org/10.1038/s41598-020-80142-2; and Lucy Plaxton, Elisabeth Hempel, William A. Marsh, Roberto Portela Miguez, Isabelle Waurick, Andrew C. Kitchener, Michael Hofreiter, Adrian M. Lister, Frank E. Zachos, and Selina Brace, "Assessing the Identity of Rare Historical Museum Specimens of the Extinct Blue Antelope (*Hippotragus leucophaeus*) Using an Ancient DNA Approach," *Mammalian Biology* 103 (2023): 549–560, https://doi.org/10.1007/s42991-023-00373-4. An older list of the bluebucks in museums is in L. C. Rookmaaker, "Additions and Revisions to the List of Species of the Extinct Blue Antelope (*Hippotragus leucophaeus*)," *Annals of the South African Museum* 102, no. 3 (1992): 131–141. For the extinction itself, see Graham I. H. Kerley, Rebecca Sims-Castley, André F. Boshoff, and Richard M. Cowling, "Extinction of the Blue Antelope *Hippotragus leucophaeus*: Modeling Predicts Non-Viable Global Population Size as the Primary Driver," *Biodiversity Conservation* 18 (2009): 3235–3242.

On the history of the Swedish natural history museum collection, see Einar Lönnberg, "Kungl. Vetenskapsakademiens Naturaliekabinett 1739–1819," in *Kungl. Vetenskapsakademien, Naturhistoriska Riksmuseets historia: Dess uppkomst och utveckling* (Stockholm: Almqvist & Wiksells, 1916), 37–38 (quote is translated by me from page 38); and Ragnar Insulander, "Det Grillska naturaliekabinettet på Söderfors," in *Anna Johanna Grill: Resedagbok från England*, ed. Catherine Lagercrantz (Stockholm: Atlantis, 1997), 153–160. For the Sparrman collection information, I have relied on Anders Sparrman, *Resa til Goda Hopps-udden, södra pol-kretsen och omkring jordklotet, samt till Hottentott-och Caffer-Landen, åren 1772–76* (Stockholm: Anders J. Nordström, 1783).

For bluebuck reports and specimens, I have used François Levaillant, *Voyage dans l'intérieur de l'Afrique, par le Cap de Bonne-Espérance, Dans les Années 1780, 81, 82, 83, 84 et 85*, vol. 1 (Paris: Leroy, 1790); Henry Lichtenstein, *Travels in Southern Africa, in the Years 1803, 1804, 1805, and 1806*, trans. Anne Plumptre (London: Henry Colburn, 1812); and Christopher I. Latrobe, *Journal of a Visit to South Africa in 1815, and 1816* (London: L. B. Seeley, 1818). The Levaillant quotes are translated by me from pages 112–113. The Latrobe quotes come from pages 168–169. I relied on the discussion of the history of bluebuck images in A. M. Husson and I. B. Holthuis, "The Earliest Figures of the Blaauwbok, *Hippotragus leucophaeus* (Pallas, 1766) and of the Greater Kudu, *Tragelaphus strepsiceros* (Pallas, 1766)," *Zoologische Mededelingen* 49, no. 5 (1975): 57–63. The image drawn by J. Allamand is found in Georges-Louis Leclerc, Comte de Buffon, *Histoire naturelle, générale et particulière: Servant de suite à l'histoire des animaux Quadrupèdes*, suppl. 4 (Schneider, Amsterdam, 1778), 151–153. The images by Levaillant are reproduced and discussed in Ian Glenn, "The Paris Bloubok (*Hippotragus leucophaeus* [Pallas, 1766] [Bovidae]) and Its Provenance," *Zoosystema* 42, no. 5 (2020): 77–84.

The Thompson index of folklore includes the "curse on treasure" as N591: S. Thompson, *Motif-Index of Folk-Literature: A Classification of Narrative Elements in Folk-tales, Ballads, Myths, Fables, Mediaeval Romances, Exempla, Fabliaux, Jest-Books, and Local Legends* (Bloomington: Indiana University Press, 1955–1958). There are some good examples of the curse in practice in John Lindow, "Swedish Legends of Buried Treasure," *Journal of American Folklore* 95, no. 377 (1982): 257–279.

For the issue of museum object repatriation and decolonization, I recommend Subhadra Das and Miranda Lowe, "Nature Read in Black and White: Decolonial Approaches to Interpreting Natural History Collections," *Journal of Natural Science Collections* 6 (2018): 4–14; Dan Hicks, *The Brutish Museums: The Benin Bronzes, Colonial Violence and Cultural Restitution* (London: Pluto Press, 2020); and Gitte Westergaard, "Colonial Entanglement in Extinction Narratives: The Afterlives of Two Saint Lucia Giant Rice Rats," *Journal of Natural Science Collections* 11 (2023): 3–12. The extinction history of the Jamaican giant galliwasp is presented in Thomas Barbour, "Notes on the Herpetology of Jamaica," *Bulletin of the Museum of Comparative Zoölogy at Harvard College, in Cambridge*, vol. 52 (1908–1910), 297. The Hunterian posted about the specimen handover on its blog: "Repatriation and New Beginnings," https://hunterian.blog/repatriation-and-new-beginnings/. On the need for a decolonial approach to environmental conservation in South Africa, see Lesley Green, *Rock |Water |Life: Ecology and Humanities for a Decolonial South Africa* (Durham, NC: Duke University Press, 2020). I heard about the "Extinction echo" concept from Sandra Swart.

CHAPTER 4

For emotional responses to extinction, I draw on Thom van Dooren, *Flight Ways: Life and Loss at the Edge of Extinction* (New York: Columbia University Press, 2014); Joshua Trey Barnett, *Mourning in the Anthropocene: Ecological Grief and Earthly Co-existence* (East Lansing: Michigan State University Press, 2022); as well as my own *Recovering Lost Species in the Modern Age: Histories of Longing and Belonging* (Cambridge, MA: MIT Press, 2019).

The first treatment of the dodo's extinction story is H. E. Strickland and A. G. Melville, *The Dodo and Its Kindred, or the History, Affinities, and Osteology of the Dodo, Solitaire, and Other Extinct Birds of the Islands Mauritius, Rodriguez, and Bourbon* (London: Reeve, Benham, and Reeve, 1848). This is the source of the Sir Hamon Lestrange story, page 22. Hilaire Belloc published his poem in 1896 in *The Bad Child's Book of Beasts* (London: Simpson, Marshall, Hamilton, Kent & Co.). The 1918 publication with "dead as a dodo" is T. Hilding Svarengren, *Intensifying Similes in English* (Lund: Lund University, 1918). See Samuel T. Turvey and Anthony S. Cheke, "Dead as a Dodo: The Fortuitous Rise to Fame of an Extinction Icon," *Historical Biology* 20, no. 2 (2008): 149–163, for a fuller treatment of the dodo's obscurity to fame story. Errol Fuller has also written significantly about the dodo in *Extinct Birds* (Oxford: Oxford University Press, 2000) and *Dodo: From Extinction to Icon* (London: Collins, 2002).

Habitat dioramas are a significant natural history display technique. In addition to the Rader and Cain book, I have used Karen Wonders, "Habitat Dioramas as Ecological Theatre," *European Review* 1, no. 3 (1993): 285–300 (quote is from page 295); and my own "Displaying Displacement: Exhibiting Extinct Birds in Natural History Museums," in *Winged Worlds: Common Spaces of Avian-Human Lives*, ed. Olga Petri and Michael Guida (London and New York: Routledge, 2023), 21–33.

CHAPTER 5

I have discussed the history of the word *endling* in Dolly Jørgensen, "Endling, the Power of the Last in an Extinction-Prone World," *Environmental Philosophy* 14 (2017): 119–138. See also Lydia Pyne, *Endling: Fables for the Anthropocene* (Minneapolis: University of Minnesota Press, 2022).

Descriptions of the passenger pigeon flocks come from John James Audubon, *Ornithological Biography, or An Account of the Habits of the Birds of the United States of America; Accompanied by Descriptions of the Objects Represented in the Work Entitled the Birds of America and Interspersed with Delineations of American Scenery and Manners*, vol. 1 (Philadelphia: Judah Dobson, 1831), with quotes from pages 320–323. The memorial commentary comes from Aldo Leopold, "On a Monument to the Pigeon," in *Silent Wings*, ed. Walter E. Scott (Madison: Wisconsin Society for Ornithology, 1947), 3–5 (the excerpt is from page 3). For Martha's current setting, see Richard Kurin, *The Smithsonian's History of America in 101 Objects* (New York: Penguin Press, 2013), chapter 28, "Martha, the Last Passenger Pigeon." I discuss more details about the passenger pigeon story in chapter 4 of my book *Recovering Lost Species in the Modern Age*.

The history of the great auk has been taken mostly from Gísli Pálsson, *The Last of Its Kind: The Search for the Great Auk and the Discovery of Extinction* (Princeton, NJ: Princeton University Press, 2024). Gísli was kind enough to share an early copy of his manuscript with me. The story of the giant anoles is from Gitte Westergaard, "Ghostly Presences: Giant Lizards on Culebra Island," in *Entire of Itself? Towards an Environmental History of Islands*, ed. Milica Prokić and Pavla Šimková (Winwick, UK: White Horse Press, 2024), 99–120.

For ecological grief, I refer to Ashelee Cunsolo and Neville Ellis, "Ecological Grief as a Mental Health Response to Climate Change–Related Loss," *Nature Climate Change* 8 (2018): 275–281; and Joshua Trey Barnett, *Mourning in the Anthropocene: Ecological Grief and Earthly Coexistence* (East Lansing: Michigan State University Press, 2022).

Referenced scientific studies of quagga are R. Higuchi, B. Bowman, M. Freiberger, O. A. Ryder, and A. C. Wildon, "DNA Sequences from the Quagga, an Extinct Member of the Horse Family," *Nature* 312, no. 5991 (1984): 282–284; J. M. Lowenstein and O. A. Ryder, "Immunological Systematics of the Extinct Quagga (Equidae)," *Experientia* 41, no. 9 (1985): 1192–1193; and M. Hofreiter, A. Caccone, R. C. Fleischer, S. Glaberman, N. Rohland, and J. A. Leonard, "A Rapid Loss of Stripes: The Evolutionary History of the Extinct Quagga," *Biology Letters* 1, no. 3 (2005): 291–295.

For the discussion of de-extinction, I have used Sandra Swart, "Zombie Zoology: The History and Reanimating Extinct Animals," in *The Historical Animal*, ed. Susan Nance (Syracuse, NY: Syracuse University Press, 2015), 54–71; Sandra Swart, "Frankenzebra: Dangerous Knowledge and the Narrative Construction of Monsters," *Journal of Literary Studies* 30, no. 4 (2014): 45–70; Adam Searle, "Exhibiting Extinction, Recovering Memory, and Contesting Uncertain Futures in the Museum," *Museum and Society* 20, no. 1 (2022): 13–32; and my own prior work on the passenger pigeon de-extinction project in *Recovering Lost Species in the Modern Age*. The Swart quotes come from page 71.

On Heck cattle as aurochs, see Clemens Driessen and Jamie Lorimer, "Back-Breeding the Aurochs: The Heck Brothers, National Socialism and Imagined Geographies for Nonhuman Lebensraum," in *Hitler's Geographies*, ed. P. Giaccaria and C. Minca (Chicago: University of Chicago Press, 2016), 138–157; Jamie Lorimer and Clemens Driessen, "Bovine Biopolitics and the Promise of Monsters in the Rewilding of Heck Cattle," *Geoforum* 48 (2013): 249–259; and taurosprogramme.com.

For more-than-human kinship, I have drawn on the ideas of Zoe Todd, "Fish Plurali-
ties: Human-Animal Relations and Sites of Engagement in Paulatuuq, Arctic Canada,"
Études/Inuit/Studies 38, no. 1–2 (2014): 217–238; and Kyle Whyte, "Indigenous Environmental Justice: Anti-Colonial Action Through Kinship," in *Environmental Justice:
Key Issues*, ed. Brendan Coolsaet (London: Routledge, 2020), 266–278.

The history of Aepyornis eggs and skeletal remains can be found in Isidore Geoffroy Saint-Hilaire, "Note additionnelle a la notice sur l'Épyornis," *Annales des sciences
naturelles*, series 3, vol. 14 (1850): 213–216; Isidore Geoffroy Saint-Hilaire, "Notice
sur des ossements et des œufs trouvés a Madagascar dans des alluvions modernes,
et provenant d'un oiseau gigantesque," *Annales des sciences naturelles*, series 3, vol. 14
(1850): 206–213; and Eric Buffetaut, "Early Illustrations of Aepyornis Eggs (1851–
1887): From Popular Science to Marco Polo's Roc Bird," *Anthropozoologica* 54, no. 1
(2019): 111–121. The H. G. Wells short story "Æpyornis Island" was reprinted in H.
G. Wells, *The Stolen Bacillus and Other Incidents* (London: Macmillan, 1904).

The huia collection information is found in K. E. Westerskov, "The Austrian
Andreas Reischek's Ornithological Exploration and Collecting in New Zealand 1877–
1889," *Otago German Studies* 1 (2016): 275–289. The sound file is in the Sound Archives
Nga Taonga, and it has been digitized. The recording can be listened to at https://www
.ngataonga.org.nz/explore-stories/stories/sound/te-karanga-a-te-huia-the-call-of-
the-huia/. The webpage also retells the history of the recording. See Julianne Lutz
Warren, "Learning a Dead Birdsong: Hopes' echoEscape.1 in 'The Place Where You
Go to Listen,'" in *Living Earth Communities: Multiple Ways of Being and Knowing*, ed.
Sam Mickey, Mary Evelyn Tucker, and John Grim (Cambridge: Open Book Pub-
lishers, 2000), 19–40, for a creative approach to the huia song.

On the gender skewing of natural history collections, see Donna Haraway,
"Teddy Bear Patriarchy: Taxidermy in the Garden of Eden, New York City, 1908–
1936," *Social Text* 11 (1984–1985): 20–64; and Rebecca Machin, "Gender Representation in the Natural History Galleries at the Manchester Museum," *Museum and
Society* 6, no. 1 (2008): 54–67.

The huia's louse was described in Eberhard Mey, "Eine neue ausgestorbene Vogel-
Ischnozere von Neuseeland, Huiacola extinctus (Insecta, Phthiraptera)," *Zoologischer
Anzeiger* 224, no. 1–2 (1990): 49–73. The specimen collected prior to extinction
held by Te Papa Tongarewa is cataloged as Extinct huia louse, *Rallicola (Huiacola)
extinctus* (Mey, 1990), AI.018365. For other extinct parasites, see Mey, "Eine neue
ausgestorbene Vogel-Ischnozere von Neuseeland"; S. V. Mironov, J. Dabert, and R.
Ehrnsberger, "Six New Feather Mite Species (Acari: Astigmata) from the Carolina
Parakeet *Conuropsis carolinensis* (Psittaciformes: Psittacidae), an extinct parrot of
North America," *Journal of Natural History* 39, no. 24 (2005): 2257–2278; and Dolly
Jørgensen, "Conservation Implications of Parasite Co-Reintroduction," *Conservation
Biology* 29, no. 2 (2014): 602–604.

The Smithsonian's acquisition of the Labrador ducks is recounted in Frederic A. Lucas, "Animals Recently Extinct or Threatened with Extermination, as Represented in the Collections of the U.S. National Museum," in *Annual Report of the Board of Regents of the Smithsonian Institution . . . for Year Ending June 30, 1889* (Washington, DC: GPO, 1891), 636. The monument to the passenger pigeon at Pigeon Hills is discussed in "Passenger Pigeon Memorial Site," *Gettysburg Times*, October 8, 1947.

On the thylacine de-extinction project and its detractors, see Amy Fletcher, "Bring 'Em Back Alive: Taming the Tasmanian Tiger Cloning Project," *Technology in Society* 30 (2008): 194–201; and Amy Fletcher, "Genuine Fakes: Cloning Extinct Species as Science and Spectacle," *Politics and the Life Sciences* 29, no. 1 (2010): 48–60.

The thylacine mother and pups display is discussed in Kathrine Medlock, "Exhibiting Extinction: Thylacines in Museum Display," *Animals, Plants and Afterimages: The Art and Science of Representing Extinction*, ed. Valérie Bienvenue and Nicholas Chare (New York and Oxford: Berghahn Books, 2022), 392–406. The article includes two photos of the display. The quote is from page 396.

CHAPTER 7

For commemoration of animals in armed conflict, see Hilda Kean, "Animals and War Memorials: Different Approaches to Commemorating the Human-Animal Relationship," in *Animals and War: Studies of Europe and North America*, ed. Ryan Hediger (Leiden: Brill, 2013); Steven Johnston, "Animals in War: Commemoration, Patriotism, Death," *Political Research Quarterly* 65, no. 2 (2012): 359–371; and Sandra Swart, "Horses in the South African War, c. 1899–1902," *Society & Animals* 18, no. 4 (2010): 348–366. Kirk Savage, *Standing Soldiers, Kneeling Slaves: Race, War, and Monument in Nineteenth-Century America* (Princeton, NJ: Princeton University Press, 1997), discusses the larger context of war memorials for the US Civil War. I have discussed the passenger pigeon memorial in Dolly Jørgensen, "After None: Memorialising Animal Species Extinction Through Monuments," in *Animals Count: How Population Size Matters in Animal-Human Relations*, ed. Nancy Cushing and Jodi Frawley (London: Routledge, 2018), 183–199.

The quotations about naming nonhumans come from Robin Wall Kimmerer, *Braiding Sweetgrass: Indigenous Wisdom, Scientific Knowledge and the Teachings of Plants* (Minneapolis: Milkweed Editions, 2013), 208; and Barnett, *Mourning in the Anthropocene*, 35.

I describe the history of the endling exhibit in Dolly Jørgensen, "Endling, the Power of the Last in an Extinction-Prone World," *Environmental Philosophy* 14, no. 1 (2017): 119–138. The bilby specimen story is recounted in Louise Boscacci, "Wit(h)nessing," *Environmental Humanities* 10, no. 1 (2018): 343–347.

The descriptions of the Carolina parakeet are from John James Audubon,

Ornithological Biography (Edinburgh: Adam Black, 1831). Quotations are from pages 135, 136, and 138.

The baiji survey that concluded it was extinct is Samuel T. Turvey et al., "First Human-Caused Extinction of a Cetacean Species?," *Biology Letters* 3, no. 5 (2007): 537–40.

The "hyperobject" concept comes from Timothy Morton, *Hyperobjects: Philosophy and Ecology after the End of the World* (Minneapolis: University of Minnesota Press, 2013).

CHAPTER 8

For thinking about play in environmental humanities, I have turned to Nicole Seymour, *Bad Environmentalism: Irony and Irreverence in the Ecological Age* (Minneapolis: University of Minnesota Press, 2018); and Alenda Cheng, *Playing Nature: Ecology in Video Games* (Minneapolis: University of Minnesota Press, 2019). I also recommend the discussion of comedy as genre in Heise, *Imagining Extinction*. On the change in natural history museums to the interactive science center model, see Karen A. Rader and Victoria E. M. Cain, "From Natural History to Science: Display and the Transformation of American Museums of Science and Nature," *Museum and Society* 6, no. 2 (2008): 152–171.

Norman MacLeod, "The Exhibition of Extinct Species: A Critique," in *Animals, Plants and Afterimages: The Art and Science of Representing Extinction*, ed. Valérie Bienvenue and Nicholas Chare (New York and Oxford: Berghahn Books, 2022), 372–391, argues that extinct species should not be exhibited in playful modes, a position I disagree with. I think Roberto Farné's position in "Pedagogy of Play," *Topoi* 24 (2005): 169–181, is a better one since he recognizes the value of play for building what he calls "political education."

Michael Blencowe, *Gone: A Search for What Remains of the World's Extinct Creatures* (London: Leaping Hare Press, 2021), chapter 3, is an easy-to-read summary of the Steller's sea cow discovery. For a more academic and thorough treatment of how the sea cow's extinction gave rise to thinking about extinction, I have used Ryan Tucker Jones, *Empire of Extinction*.

The *REViVRE* experience was made by Saola Studio and is described on the company's website: https://www.saolastudio.com/en-gb/revivre. For a good overview of the *Skin & Bones* application and the thinking behind it, see Diana Marques and Robert Costello, "Skin & Bones: An Artistic Repair of a Science Exhibition by a Mobile App," *MIDAS* 5 (2015), doi:10.4000/midas.933. *Dodo Expedition AR* is discussed in Mauritius Museums Council, "Dodo Augmented Reality Experience," https://mauritiusmuseums.govmu.org/mauritiusmuseums/?p=4115.

For scholarship on sound recordings for conservation of birds and to document extinction (or not), see Alexandra Hui, "Listening to Extinction: Early Conservation Radio Sounds and the Silences of Species," *American Historical Review* (2021): 1370–1395; and Joeri Bruyninckx, *Listening in the Field: Recording and the Science of Birdsong* (Cambridge, MA: MIT Press, 2018). The Hawaiian capes made from ʻōʻō feathers are discussed in Gitte Westergaard, "Hidden Stories of Extinction: Hawaiian ʻAhuʻula Feather Capes as Biocultural Artefacts," *Museum and Society* 20, no. 1 (2022): 104–117.

The Clifton Park Museum archives has some newspaper clipping archives that include articles with Nelson. Thanks to the curators for showing them to me. The quote related to the sixty-fifth-anniversary coverage comes from "Milestone for Lion Who Is King of Museum," *Yorkshire Post*, https://www.yorkshirepost.co.uk/news/milestone-lion-who-king-museum-1931442.

For a good overview of how to design environmental education for environmental citizenship, see Andreas Ch. Hadjichambis and Demetra Paraskeva-Hadjichambi, "Education for Environmental Citizenship: The Pedagogical Approach," in *Conceptualizing Environmental Citizenship for 21st Century Education*, ed. A. C. Hadjichambis et al. (Cham: Springer, 2020), 237–61.

For craftivism and environment, see Sarah Wade, "The Art and Craftivism of Exhibiting Species and Habitat Loss in Natural History Museums," *Museum and Society* 20, no. 1 (2022): 131–146. Brandon Ballengée's artist statement is on his website: https://brandonballengee.com/the-frameworks-of-absence/. The comment from Valérie Bienvenue and Nicholas Chare is in "After Extinction," in *Animals, Plants and Afterimages: The Art and Science of Representing Extinction*, ed. Valérie Bienvenue and Nicholas Chare, (New York and Oxford: Berghahn Books, 2022), 407–425, on page 408. Reflections on Audubon and capitalism are found in Gordon M. Sayre, "Rare Birds and Rare Books: The Species as Works of Art," in *Animals, Plants and Afterimages: The Art and Science of Representing Extinction*, ed. Valérie Bienvenue and Nicholas Chare (New York and Oxford: Berghahn Books, 2022), 191–210.

The details about the Smithsonian thylacine were found at http://thylacine.psu.edu/history.html and https://web.archive.org/web/20180912155532/https://naturalhistory.si.edu/onehundredyears/featured_objects/Thylacine.html.

CHAPTER 9

I have written about the Caribbean monk seal's history and specimens collected by museums in Dolly Jørgensen, "Erasing the Extinct: The Hunt for Caribbean Monk Seals and Museum Collection Practices," *História, Ciências, Saúde-Manguinhos* 28, suppl. (2021): 161–183. The first specimen on display is discussed in Joel Asaph Allen, "The West Indian Seal (*Monachus tropicalis* Gray)," *Bulletin of the American Museum of Natural History* 2 (1890): 1–34. The Mexican specimen that was lost is discussed in Mario Irepan Luna-Pérez and Consuelo Cuevas-Cardona, "La extinción de la Foca Monje del Caribe," *Herreriana* 3, no. 2 (2022): 12–17. The mention of the Harvard Caribbean monk seal appears in Harvard MCZ, *Notes Concerning the History and Contents of the Museum of Comparative Zoology* (Cambridge, MA: Harvard University, 1936), 42. I found the photograph of the specimen previously on display at https://dinopedia.fandom.com/wiki/Caribbean_monk_seal.

I have relied on several scientific studies of Delcourt's gecko in my discussion of its history: Aaron M. Bauer and Anthony P. Russell, "*Hoplodactylus delcourti* n. sp. (Reptilia: Gekkonidae), the Largest Known Gecko," *New Zealand Journal of Zoology* 13, no. 1 (1986): 141–148, which is the first description of the animal; Darren Naish, "New Zealand's Giant Gecko: A Review of Current Knowledge of *Hoplodactylus delcourti* and the Kawekaweau of Legend," *The Cryptozoology Review* 4, no. 2 (2004); and Matthew P. Heinicke, Stuart V. Nielsen, Aaron M. Bauer, Ryan Kelly, Anthony J. Geneva, Juan D. Daza, Shannon E. Keating, and Tony Gamble, "Reappraising the Evolutionary History of the Largest Known Gecko, the Presumably Extinct *Hoplodactylus delcourti*, via High-Throughput Sequencing of Archival DNA," *Scientific Reports* 13 (2023): no. 9141, https://doi.org/10.1038/s41598-023-35210-8. The children's book is Laurence Talairach, *La malédiction du gecko* (Toulouse: Plume de carotte, 2019).

The list of giant earwig specimens is taken from the Global Biodiversity Information Facility database (gbif.org). On the rice rats, see Westergaard, "Colonial Entanglements in Extinction Narratives." Information on the small Mauritian flying fox, along with the text of the letter from 1772 about the flying foxes on Reunion Island, is from Anthony Cheke and Julian Hume, *Lost Land of the Dodo: An Ecological History of Mauritius, Réunion & Rodrigues* (London: T. & A. D. Poyser, 2008).

Transcriptions of the letters about the Xerces blue butterfly were published in F. Martin Brown, "Letters from Dr. H. H. Behr to Herman Strecker," *Journal of the Lepidopterists' Society* 22, no. 1 (1968): 57–62. For contemporary genomic research on the Xerces blue that confirms it as a distinct species, see F. Grewe, M. R. Kronforst, N. E. Pierce, and C. S. Moreau, "Museum Genomics Reveals the Xerces Blue Butterfly (*Glaucopsyche xerces*) Was a Distinct Species Driven to Extinction," *Biology Letters* 17 (2021): article 20210123, https://doi.org/10.1098/rsbl.2021.0123; and Toni de-Dios, Claudia Fontsere, Pere Renom, Josefin Stiller, Laia Llovera, Marcela Uliano-Silva, Alejandro Sánchez-Gracia, Charlotte Wright, Esther Lizano, Berta Caballero, Arcadi Navarro, Sergi Civit, Robert K. Robbins, Mark Blaxter, Tomàs Marquès-Bonet, Roger Vila, and Carles Lalueza-Fox, "Whole-Genomes from the Extinct Xerces Blue Butterfly Can Help Identify Declining Insect Species," *eLife* 12 (2023): RP87928, https://doi.org/10.7554/eLife.87928.1. There is an eloquent description of the Xerces blue's extinction and the author's encounter with specimens in the back room in Blencowe, *Gone*, chapter 7.

I highly recommend Thom van Dooren, *A World in a Shell: Snail Stories for a Time of Extinctions* (Cambridge, MA: MIT Press, 2022), for insights into the past and ongoing extinctions of Hawaiian snails.

For up-to-date research on the Japanese sea lion's population history and extinction, see Yoon-Ji Lee, Giphil Cho, Sangil Kim, Inseo Hwang, Seong-Oh Im, Hye-Min Park, Na-Yeong Kim, Myung-Joon Kim, Dasom Lee, Seok-Nam Kwak, et al., "The First Population Simulation for the *Zalophus japonicus* (Otariidae: Sea Lions) on Dokdo, Korea," *Journal of Marine Science and Engineering* 10, no. 2 (2022): 271, https://doi.org/10.3390/jmse10020271. I have given both the Korean and Japanese names for the island because it is politically contested geography by the two nations. The animal is generally referred to as Japanese sea lion in scientific literature, but I note that Korean authors prefer to call it Dokdo sea lion.

EPILOGUE

On the function of memorial museums, see Amy Sodaro, *Exhibiting Atrocity: Memorial Museums and the Politics of Past Violence* (New Brunswick, NJ: Rutgers University Press, 2018). The argument about slowing time in museums is found in Dolly Jørgensen, Libby Robin, and Marie-Theres Fojuth, "Slowing Time in the Museum in a Period of Rapid Extinction," *Museum and Society* 20, no. 1 (2022): 1–12.

Index

Page numbers in italics refer to figures.